Holography Projects
for the Evil Genius®

Evil Genius® Series

Holography Projects
for the Evil Genius®

Gavin D. J. Harper

New York   Chicago   San Francisco   Lisbon   London   Madrid
Mexico City   Milan   New Delhi   San Juan   Seoul
Singapore   Sydney   Toronto

Copyright © 2010 by The McGraw-Hill Companies, Inc. All rights reserved. Printed in the United States of America. Except as permitted under the United States Copyright Act of 1976, no part of this publication may be reproduced or distributed in any form or by any means, or stored in a data base or retrieval system, without the prior written permission of the publisher.

1 2 3 4 5 6 7 8 9 0   WDQ/WDQ   1 6 5 4 3 2 1 0

ISBN 978-0-07-162400-8
MHID 0-07-162400-7

Sponsoring Editor	**Copy Editor**
Judy Bass	Anne Lesser
Editing Supervisor	**Proofreader**
Stephen M. Smith	Medha Joshi, Glyph International
Production Supervisor	**Indexer**
Pamela A. Pelton	Robert Swanson
Acquisitions Coordinator	**Art Director, Cover**
Michael Mulcahy	Jeff Weeks
Project Managers	**Composition**
Smita Rajan and Vasundhara Sawhney, Glyph International	Glyph International

Printed and bound by Worldcolor/Dubuque.

McGraw-Hill books are available at special quantity discounts to use as premiums and sales promotions, or for use in corporate training programs. To contact a representative, please e-mail us at bulksales@mcgraw-hill.com.

This book is printed on acid-free paper.

To Missy the cat, for keeping the PC chair warm when I wasn't there.
And for providing me with endless amusement while I edited the final drafts
of this book, by chasing the chess pieces and optical mounts around the floor while
I was trying to write and interrupting my flow by bringing me "presents" to dispose of.

About the Author

Gavin D. J. Harper is the author of several popular books on various science and technology subjects. Described by *The Independent* as a "whizz-kid on a mission," his work has been cited in sources as diverse as *Science, MAKE,* *The Ecologist*, and *The Inquirer*. He also writes regularly for *Green Building Magazine* and *The Huffington Post* blog. He lectures regularly throughout Europe and is passionate about communicating science and technology subjects, working with the Royal Institution's Engineering Masterclasses program to enthuse young people. Gavin is a Member of the Institute of Engineering and Technology. He holds a Diploma in Design & Innovation, a BSc (Hons) in Technology, and a BEng (Hons) in Engineering from the Open University. He holds the Diploma of Vilnius University, Lithuania. He went on to study toward an MSc in Architecture with Advanced Environmental and Energy Studies from the University of East London at the Centre for Alternative Technology and holds an MSc in Social Science Research. Gavin is currently reading for his Ph.D.

 Gavin is pictured next to an "i-Lumogram" digital hologram, prepared by Geola UAB, Vilnius, Lithuania. The hologram is created digitally from a sequence of images recorded continuously as a video. For more information, check out www.geola.lt/lt/digital_holography/.

Contents

Acknowledgments

First of all, a great thank you to Dr. Andrew Norton of the Open University, Milton Keynes. He "lit the touch paper," so to speak, by helping me to procure my first lasers and optics equipment and finding me a copy of the course materials from OU course ST291, "An Introduction to Holography." It's yet another example of how I'm indebted to the Open University, a great institution, for encouraging me along the way.

Thanks also to Rob Munday and Patrick Boyd for encouraging me during the early stages of writing this book.

Immense thanks to Jeff Blyth for his unceasing encouragement via Skype and for sharing his wisdom. A few sessions with Jeff and my ideas finally began to become coherent.

I'd also like to thank Alec Jeong, son of Tung H. Jeong, whose name you will find mentioned in this book. As managing director of Integraf (listed in the Suppliers' Index), Alec was great at getting me holography materials shipped across the Atlantic when I needed them yesterday. A great family business, Integraf is speedy and reliable for supplying holography materials. A big thank you also to Randy of Industrial Fiber Optics for his swift service.

The holography movement is reliant on a group of individuals moving the hobby along one step at a time in small increments. Formulas for new emulsions and development processes do not just spew forth from large chemical manufacturers. Rather, they are the product of trial and error, countless experiments, and dedication by a small band of dedicated people. New techniques need to be developed and tested, and in such a small community advanced hobbyists are at the vanguard of this change.

Thanks also to Alan Stein, who devised the smashing little application and has been a mine of knowledge in helping me understand computer-generated holography; it's really great when people create a tool and make it open source so everyone can learn.

My thanks to Judy Bass, a wonderful editor, for having the patience of two saints and the good nature to bear with me until the very end.

Lastly, "Ačiū gausa" to my new friends in Vilnius, Lithuania—Dr. Stanislovas Zacharovas, Jevgenij Kuchin, and Giedrius Gudaitis of Geola—for producing the "i-Lumogram" integral hologram, pictured in the About the Author section, from a video filmed by my unceasingly fabulous colleague, Emma Dean.

Disclaimer

In this book we are going to be working with a variety of chemicals, some of them really benign, others particularly nasty, which have the possibility (depending on the chemical) to burn you, blind you, knock you out with noxious vapors, make you ill—or worse.

In preparing this book, I've tried to alert you wherever we will be using something particularly nasty. However, if you're going to be carrying out any of the great projects in this book, and you aren't a responsible adult, go find one now and show them this.

Note to Responsible Adult

Science is *really* good fun, and no one ever got anywhere without taking a measured amount of risk. Scientists of old had lab assistants who were pretty expendable, if they lost an eye or appendage, or even died. But it is your responsibility to ensure the small person you are responsible for is safe and doesn't come to any harm.

Goggles

At all times when working with lasers, I highly recommend the use of laser safety goggles. You need to choose a set of goggles with a lens color that is the "opposite" of the color of laser that you will be working with. So if you are working with blue or green lasers, then pick goggles with red lenses, and if you are using red lasers, pick goggles with blue lenses. To make matters more complicated, when you are dealing with chemicals,

make sure you find some goggles that are splash-proof and will protect your eyes from any spillages and mishaps! The following sign has been doing the rounds on the Internet. Although humorous, it also sounds a cautionary note!

Big Scary Laser
Do not look into beam
with remaining eye

Chemical Safety

With the advent of the Internet, it is now easier to find out information about chemical safety than ever before. A quick typing of a chemical name into a good search engine, followed by "MSDS," should yield a plethora of Material Safety Data Sheets. Know thy enemy! These sheets might not be the most gripping read in the world, but a small investment of time is going to ensure that you are safe, can allow for any risks, and have taken action to mitigate any danger to which you might be exposing yourself.

Protecting Your Skin

Some of the chemicals we are going to use in this book are particularly nasty, and some of the bleaches may cause skin burns or irritation. So if you are going to be working with these chemicals, cover up. Wear gloves to protect your hands, and wear long-sleeved (preferably thick) clothes.

Depending on who you ask, holography is something different. To the scientist, holograms form as the result of interactions between coherent beams of laser light on an emulsion coated with silver halide compounds. To the artist, they are objects of great beauty, transcending simple visual illusions of depth and providing a medium where three dimensions can truly be recorded on a flat two-dimensional plane. However, although perspectives and viewpoints differ, the one thing that unifies creators and viewers of holograms is that they are exciting!

Holography is a great hobby because it allows you to create something that will amaze and wow while at the same time learn all sorts of science. It's also good because it bridges what some perceive to be the immutable divide between creative and science subjects. Holography affords artists a plethora of creative new artistic opportunities, enabling them to escape the planar constraints of the canvas and jump out into the third dimension. For scientists, it provides a rich vein for scientific inquiry and experimentation.

This book is a gentle introduction to the art and science of holography. Chapter 1 provides a very brief historical orientation to holography, showing its relatively recent origins and the developments that have taken place in the past few decades. In Chap. 2, we cover the basics of human vision, how we see and how we interpret three-dimensional images, which is key to understanding how three-dimensional imaging works. In Chap. 3, we provide a run-through of basic optics but done in a way that is practical. By the time you've finished this chapter, you should understand how we're going to bend and distort laser light to form a hologram, have a grasp of most basic optical components, and also understand some of the

practical aspects of working with these. Chapter 4 provides an orientation to the physics of light and the special properties of lasers that make holography possible. It also provides practical information regarding the selection of lasers for holography.

By the time we reach Chap. 5, we should be able to unify all of this knowledge into an explanation of the physics of how holography actually works, understanding what makes holograms special and different from other three-dimensional imaging technologies. When we understand the process of holography, we should move on to Chap. 6, which explains what goes on with the photosensitive emulsion that records our hologram. We also look at the basic method to process a hologram using off-the-shelf chemical mixes. This chapter covers the fundamentals and basics of holographic chemistry, and these principles will apply to the holograms that you produce in the next few chapters. Chapter 7 covers the fundamental equipment you should consider buying and making for your holography workshop, showing items of equipment that you should purchase and items you can make.

In Chap. 8, we'll be looking at simple holographic projects. These require the minimum of optics and equipment and should be quite simple to produce for the beginner. If you want to just say, "I've made a hologram," by the time you finish this chapter, your objective should be accomplished! For the simpler holograms, we're going to stick with using glass plates, which although a little more expensive are simpler to work with for the beginner. If you want to take things further, however, read on to Chap. 9, which tackles some slightly harder holography projects. In this chapter we explore how you can swap out the glass plates in your projects for film, which works out cheaper but is trickier to work with, and we cover some more advanced setups. If you work through these projects successfully, you should be well on your way to becoming a holography pro!

Chapter 10 covers more advanced holographic setups that increase in their complexity and require use of a greater array of optical components and a bit more skill in setup. Nevertheless, if you can master these setups, you will find they produce very rewarding results! Chapter 11 is one for the chemists; it covers the preparation of some more-advanced holographic chemistry, which if you have access to chemicals through a college or university can produce different and more-varied results. Chapter 12 looks at the potential for using computers to generate holograms. Using a freely available program, you should be able to experiment with very simple computer-generated holograms that can be printed on a good-quality black-and-white laser printer. Chapter 13 is one for the electronics buffs and covers the construction of a handful of electronic circuits that are useful to budding holographers. They aren't essential to the projects in the book, but you may find that if you are making a lot of holograms, some of these are very useful.

Chapter 14 provides some fodder for science fair projects. If you want to look at some different applications of holography, you should be able to take the principles you have learned in this book and apply them to developing some of the suggested themes in this chapter. By the time we get to Chap. 15, the holography is nearly over! People use the word hologram in everyday usage to refer to a variety of different three-dimensional imaging techniques that do not technically qualify as holography. In this chapter, we debunk some of this by looking at a variety of different technologies that although they do not qualify as "holograms" are interesting in their own right and worth a brief moment of exploration. Chapter 16 is the "what next" part of the book. As a future holographer, armed with the knowledge in this book and perhaps some future learning, you will be developing the next (coherent) wave of holographic innovations!

As we look forward into the future, though, it's important to take a brief look back at where the hobby has come from. The technology is only just in its sixth decade. This book comes at a rather historic moment, with the first printing scheduled for June 2010. The first copies of this book will be off the printing presses in time to celebrate the 50th birthday of the first practical demonstration of the laser, the enabling technology that has made holography of the type we would recognize possible.

I hope you enjoy working through some of the projects in this book. If it gets you hooked, this is only the start of the journey!

Gavin D. J. Harper

Holography Projects
for the Evil Genius®

Chapter 1

History of Holography

Holography is now almost ubiquitous. From the little images on our credit cards, to the security seals on genuine DVD cases, to promotional items, holography is now all around us. When your goods are scanned at the supermarket checkout, you are probably unaware of the small holographic optical element turning a single laser beam into the spirograph patterns that scan your bar codes, but this has not always been the case.

The history of holography is a relatively recent one, but the roots of the word are old. The roots of the words *hologram* and *holography* are from Greek; *holos* means "whole," and the "gram" ending signifies the Greek word for "message." The word had been used in a limited capacity before Denis Gabor's reinvention of the term; some Old English dictionaries refer to a "hologram" as an important letter handwritten and signed by the same person.

Although this isn't an exhaustive history of holography, it should provide some orientation to how far the hobby and technology has come in such a short space of time, and it should aid and inform your understanding of the cultural history of the experiments you are about to undertake.

Gabor was working on trying to improve the process of electron microscopy when he discovered the principles of holography in 1947.

Figure 1-1 shows the original setup that Denis Gabor used. Because the laser had not yet been invented, the only source of monochromatic light that Gabor could use was a mercury arc lamp. The light produced by these lamps is far inferior to laser light; as a result, Gabor's early holograms were of poor quality. Furthermore, additional defects were introduced by the fact that all of the

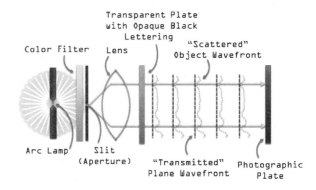

Figure 1-1 *Gabor's original holography setup.*

optical elements were in the same line, or axis. As we will see with the single-beam holograms we are going to produce, an "off-axis" setup produces far superior results.

From Gabor's original invention, the field didn't really develop for a decade. There just wasn't a technology available that could deliver the special quality of light required to make good quality holograms. We had to wait for the invention of the laser for holography to really take off.

The first hologram was a series of names of scientists spelled in capital letters: NEWTON, HUYGENS, YOUNG, FRESNEL, FARADAY, MAXWELL, KIRCHHOFF, PLANCK, EINSTEIN, and BOHR. We can see Gabor's original image in Fig. 1-2, the image produced by the hologram, and the reconstructed image using the light from the mercury arc lamp. Notice the artifacts on the hologram.

The main innovation that has enabled the modern holography we take for granted today is the laser. As this book goes to press in May 2010, it will be exactly 50 years since the first practical demonstration of laser technology by Theodore Maiman of Hughes Laboratories.

| Original Image | Hologram | Reconstructed Image |

Figure 1-2 *Gabor's source image, hologram, and reconstructed image.*

The idea behind the laser was formulated simultaneously on opposite sides of the world by Charles Towns and Arthur Schawlow in the United States and by Aleksandr Prokhorov and Nikolay Basov in the former Soviet Union, now part of present-day Russia. They both devised the concept of an open-cavity resonator, unaware of each others' work. Originally Towns and Schawlow were concentrating on producing microwaves in a device they called a "maser"; however, as noted in the box feature, the development of the laser, a resonator producing an optical output, is shrouded in controversy and was contested 28 years later by Gordon Gould.

The laser bought a light source that was not only stable and not prone to fluctuations or deviations in the light output, but it also had a high "coherence length," a concept explained more thoroughly later in the book.

In Maiman's experiment, a rod of ruby was stimulated by bright flash bulbs. The quality of light produced by this laser is fundamentally different from the continuous beams produced by the helium neon or diode lasers you will be using in this book. Rather than a continuous steady beam, the laser that Maiman developed is known as a "pulsed laser." We don't use them here because they are expensive and beyond the reach

The word *laser* has mutated from what was originally the acronym LASER, which stood for "light amplification by the stimulated emission of radiation." The key concepts behind laser technology were developed as early as 1917 when Albert Einstein theorized about the process of "stimulated emission." The word *laser* is shrouded in controversy; Gordon Gould is the first person on record to use the word. He studied under Charles Townes, who invented the maser, predecessor to the laser (switch the word "light" for "microwave," and you'll understand). Gould believed that the earlier masers, which produced microwaves rather than visual light, could be modified to produce an optical output. He started to build his laser in 1958, but he didn't file any papers for the invention until later in 1959. Alas, by this point he had been beaten to the punch by his supervisor, and his patent was refused. A long and bitter patent war ensued, and it was not until 1977 that the patent for the laser was finally granted. Ironically, this worked out better for Gould in the long run. By the time he eventually got his patent, many applications had been developed for the laser and their use was becoming commonplace in many spheres of life. Because the patent is granted for a fixed period, Gould made more money by waiting than he would have had his patent been granted when he first applied.

History of Holography

of the beginner holographer; however, it is good to be aware of their existence because the short burst of intense light that only lasts a few nanoseconds is ideal when it comes to capturing subjects that are prone to move—for example, if making a holographic portrait of a person. On that note, it is interesting that the first holographic portrait was made in 1967.

However, there is still a bit of work to be done before we get to 1967. The technology of the laser still needed to develop a little before then. So the ruby lasers of Maiman were all well and good, but they required rods of synthetic ruby, which is expensive, and they only produce brief pulses of light. It wasn't until the end of 1960 (in a work published in 1961) that Ali Javan, William R. Bennett Jr., and Donald Herriot designed the helium neon laser. This was the first gas-discharge laser, and it is a common sight on many a holographer's workbench. It had a couple of unique identifying traits. First, as we have seen, it produces a continuous output of laser light, and second, it doesn't require "flash bulbs" to optically stimulate it. Instead it uses an electrical discharge to excite the gas. That's enough about lasers. Let's get back to the holography story, for now at least.

Two scientists working at the University of Michigan, Emmett Leith and Juris Upatnieks, having read Denis Gabor's paper, decided they were going to replicate his work. However, they would change it subtly by borrowing a technique they had used in working with side-reading radar. Rather than having all of the components in the same line as Gabor had done, they moved to an "off-axis" setup, the kind you will see later in this book, to make a transmission hologram of a toy train and bird. The off-axis technique is recognized by modern holographers and explored here.

We are still at the point in history where holograms required laser light in order to replay the effect. Meanwhile, at the other side of the world, Yuri N. Denisyuk, working in the former Soviet Union, took ideas from holography and combined them with work done by Gabriel Lippmann into a technique for color photography. Where Denisyuk's holograms differed is that, unlike Leith and Upatniek's holograms, they were *reflection holograms* rather than *transmission holograms.* Furthermore, his holograms could be viewed in white light, with no need for a laser to replay them.

In the same year, Robert N. Hall, based at a General Electric research center, made the discovery of coherent light emission from a semiconductor diode. A little later that year, Nick Holonyak Jr. (with a name like that he was bound to end up written into the history of holography!) demonstrated the first visible wavelength laser diode. However, there was the small matter that these lasers again only produced a pulsed output, and, even worse, required cooling to 77 kelvins (K)—liquid nitrogen temperatures that would have put them well out of the reach of the average holographer.

In 1968, Stephen Benton discovered the rainbow transmission hologram. This was the first transmission hologram that could be viewed in white light. As the viewers changed their perspective, the hologram changed through the colors of the rainbow. Previously if you had tried to view a transmission hologram using white light, you would have seen a rainbow-colored blur but no sharply defined image. This was one of the crucial innovations to bring holography to the masses and enabled the invention of the embossed hologram.

We now commonly see holograms "embossed" in Mylar. Although it is beyond the scope of this book to show how to produce high-volume commercial holograms, it is an important discovery. Effectively, a metal "stamp" is made from a holographic master. This stamp takes the form of a drum. A sheet of Mylar is pinched between the drum and a roller so that as the Mylar passes through, the image of the master is stamped onto the holographic foil. What results is a cheap hologram that can be rapidly reproduced using simple materials. We now find these embossed holograms on a variety of objects, from credit cards to designer wear, often acting as a security function to verify authenticity.

Lloyd Cross helped the hobby make a great leap forward by inventing the simple sand table system that we will show you how to construct in this book. The sand table does away with expensive precision optical components and replaces them with a simple sand pit and cheap holders for optical components made from a variety of readily available materials.

Meanwhile, back to the laser lab, interesting things were happening on opposite sides of the globe. Zhores Alferov in the Soviet Union and partners Izuo Hayashi and Morton Panish both managed to produce laser diodes that would produce a continuous beam and work at room temperature. Since their invention, these devices have steadily infiltrated all manner of electrical equipment from bar code scanners to blue ray video players. And their price has come down dramatically to the point where they now represent one of the most cost-effective routes into amateur holography.

In 1971, as the value of holography was starting to be realized through a variety of different applications, the work of Denis Gabor was honored with a Nobel Prize for his discovery of holography.

Later acquired in 1993 by the Massachusetts Institute of Technology where it now resides, the Museum of Holography was founded in 1976 in New York.

Holograms on credit cards to verify authenticity are now a ubiquitous sight. We take them for granted; however, how many of us stop to think that it was only as recently as 1983 when MasterCard International Inc. became the first bank to use holograms as part of their card security?

In March 1984, *National Geographic* was the first large-scale circulation magazine to carry a hologram embossed on the cover. Then, in December 1988, *National Geographic* went for the big time: a whole magazine cover composed of an embossed hologram. As far as I know, no magazine publisher has ever produced a wholly holographic magazine, although it's worth throwing down the gauntlet to an enterprising publisher!

Since then, there have been a number of advances, notably in the field of digital photography, using computers to generate and master holograms. We still have a long way to go until we realize the dreams of science fiction, but we'll save those for the final chapter that discusses the future of holography.

Until then, work your way through some of the projects in the book, and see how far the technology has come. If the field of holography catches you, who knows? You could even become the pioneer of the next major holographic innovation!

Chapter 2

How We See in Three Dimensions

What distinguishes holography from many other forms of visual media and makes it special and unique is its ability to record things in three dimensions. With the advent of computer graphics, games consoles, and the advanced rendering capabilities that have been made available by the rapid advances in microprocessor technology, we've almost come to take three-dimensional (3D) imagery for granted.

But although it is possible to make 3D renderings of objects or computer-generate imagery, capturing an actual object with photographic quality and having the ability to view it from multiple angles is something really quite special indeed.

Putting the problem of 3D imaging into perspective, the human race likes to record information about important things, objects, events, and people. Since the days of cave paintings, we have found increasingly elaborate methods to record our experiences using a variety of different media. With the development of painting and the quest for realism to the development of photography, technology has enabled gradual improvements in the quality of information we are able to record.

But to understand how we can record 3D information, we first need to take a step back to understand how we perceive the world in this way.

There are well-composed pictures that although do not give us true 3D information, exhibit the property of depth that allows us to perceive what images are near and what ones are far. How do we do this?

Well, the science behind understanding perspective is really easy to understand. Without delving into the anatomy of the eye too deeply (find a good biology textbook), we'll assume you understand that images enter the eye and are formed upside down on the retina at the back of the eye. The brain takes the information coming from the eye and makes sense of it to form the scene in our brains.

Now, take an object of set size. When the object is nearby, the "solid angle" of our field of vision that object occupies is large; it takes up a lot of space in our view because it forms a large image on our retina. Move further away from the object or move the object further away from us, and the size of the retinal image decreases. See Fig. 2-1.

This can have comic effects. We've all seen the holiday snapshots that are badly taken, where the Eiffel Tower, poorly positioned in the frame, ends up looking like a hat on a relative's head. But because our brain has an idea of the scale of the Eiffel Tower (we know it's really big), our brain comes to the conclusion that it is a "big object in

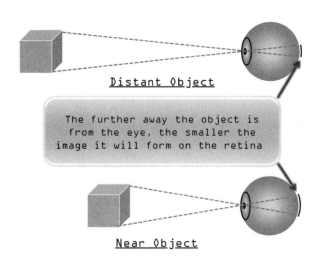

Figure 2-1 *How our eyes perceive perspective.*

the distance." So the thing that gives us the visual clues is the *relative* retinal image size.

When we look at a picture, the plane of the paper or canvas is a fixed distance away from the viewer, yet some objects can be made to seem in the distance while others seem nearby. In a two-dimensional (2D) image, the only way we can achieve this effect is with perspective depth cues, making things smaller or bigger relative to each other and putting things in front of or behind each other. But with holography, we have a much richer range of depth cues.

It's interesting that when looking at the world of art, attempts at visual realism in three dimensions are relatively recent. In fact, when the concept of perspective was discovered, so enthralled were the early pioneers that they formed a group in northern Italy called "i Perspectivi."

M. C. Escher was one artist who mastered the art of depth and perspective to the point where he could create impossible 3D structures mapped onto a 2D woodcut. Put his name into your search engine's image search, and you'll yield a bewildering array of impossible structures that, although feasible on the page, could never translate into a 3D form. You can focus on parts of the image, and the sections on their own seem "plausible" in three dimensions. But these components are configured, such that when assembled together into a composite image, they couldn't possibly work in real life.

The perspective cues can make 2D images seem three dimensional, but what about those cues that are exclusive to 3D viewing? Think about a camera and the human head: What is the difference? One of the differences is that a camera captures a scene using a single lens, whereas our heads contain two eyes that capture two slightly different viewpoints that the brain must fuse together. This is quite important because depending on how far away objects are, the "angular separation" between the line of sight of your two eyes changes. You can prove this to yourself. Hold your finger upright as far away

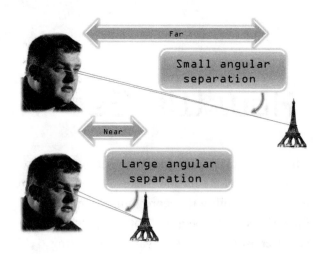

Figure 2-2 *How angular separation changes with distance.*

from you as possible, and concentrate on the end. Now slowly move the tip of your finger toward your nose, and concentrate on focusing on the tip of it. What happens by the time that your fingertip touches your nose? Your two eyes are pointing inward and focusing on your fingertip. *The angular separation has changed as you moved your fingers.* We can see this concept in Fig. 2-2.

We call the fact that we see with two eyes *binocular vision* in stereo photography, which you will experiment with in the chapter on other 3D projects that aren't "holography." You'll see how by increasing the information we capture, by using two cameras rather than one, we can capture enough information to give our eyes and brain sufficient data to replay the image in three dimensions with some clever viewing methods that "fool" our eyes and brain into thinking that they are seeing a 3D image. *But this is quite different from holography!* When we produce a hologram, we are capturing the actual wave fronts of light waves reflected from an object. When we view a hologram, we are replaying *an actual 3D image.* When we change our viewpoint, the image moves as it would in real life. This is the distinguishing feature that differentiates holography from other 3D photo recording methods.

Other cues work to help us form a 3D image. If we place one object in front of another, it obscures the object behind and so we may only see part of it; this helps our brain to form the picture of where the objects reside in 3D space. In 2D imagery, we can only see a 3D perspective from a static viewpoint; objects that are obscured will always remain obscured. However, the beauty of the art of holography is that viewers are able to change their position. With a little cunning, the holographer can devise 3D schemes in which one minute you see an object, but by changing your perspective the

object becomes obscured by another in the scene. This is where holography comes into its own and is unmatched in its ability to show a collection of objects from a diverse range of viewpoints.

So what makes a hologram different from a conventional picture? Well, to understand this, it's probably easier to get some understanding of conventional pictures, and we're going to do this in a practical way by exploring the camera obscura, also known as a "pinhole camera."

Project 1: Make a Camera Obscura

You Will Need

- Black card
- Tracing paper
- Aluminium foil
- Glue
- Black electrician's tape
- Large sheet of heavy black material or towel

Tools

- Scissors
- A pin

You can demonstrate the principles behind a pinhole camera very simply. The essence of the method is this: A dark box contains a pinhole at one end and a translucent screen on the other. The pinhole will help focus light from the incoming scene onto the screen or film at the rear of the camera (as shown in Fig. 2-3).

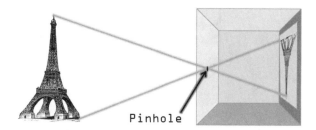

Pinhole

Figure 2-3 *How a pinhole camera works.*

Construct a simple box out of black card (a suggested net is shown in Fig. 2-4; at the front of the box, you'll see a rather large hole. Stick a sheet of aluminium foil over this hole, and using a fine pin, punch a hole in the center. To the rear of the box, use a sheet of greaseproof or tracing paper to act as a translucent screen onto which you can project the image formed by the pinhole.

You can then fold the net, and using some glue, stick the flaps to the inside of their adjacent card sides when folded to form a light tight box. If there are any gaps or cracks into which light could enter the box, other than the pinhole or translucent screen, take some opaque black electrician's tape and cover them over.

Now take your box outside and find a brightly lit scene to "capture" with your pinhole camera.

One of the things you will notice is that the image formed on the back screen is relatively dim, to the

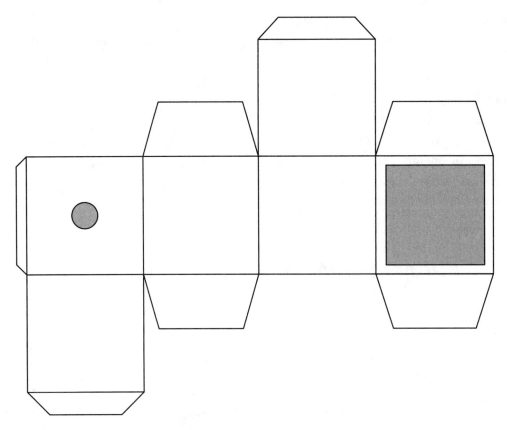

Figure 2-4 *Template for a pinhole camera.*

point where the ambient light has the potential to overwhelm the image on the screen. For this reason we have the sheet of dark fabric. If you put this over your head, and wrap it around the camera in front of you, making sure the pinhole is still exposed to the outside, you should be able to exclude outside light. Your eyes will then adjust to the darkness, and the image on the screen will seem brighter and sharper.

You can take this simple camera obscura and with the aid of some black-and-white photographic paper and chemicals, you can produce photographs by replacing the tracing paper with a sheet of photographic paper with the emulsion side facing inside the box. Tape well to ensure no stray light can get in, and cover the pinhole until you are ready to take your picture. Go outside, find somewhere still, and uncover your pinhole. You will need to experiment with different exposure times, which may run to around 2 min or so.

When the image has been exposed, develop the print in black-and-white photo chemistry. Some of the better photographic shops will sell this paper and chemicals. The process is broadly similar to developing holograms, which you will read about in Chap. 6. We can't go into too much information here, but there are some good resources on the Internet if you want to explore this further as a bonus project. If you are preparing a "lab" to process your holograms, then you will have most of the equipment, and it is worth the small investment to try this out.

www.instructables.com/id/How_To_Make_A _Pinhole_Camera/www.youtube.com/watch?v =KmJznKe4jpI

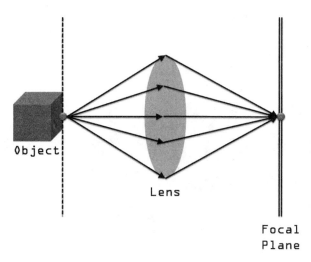

Figure 2-5 *A lens produces a focused image.*

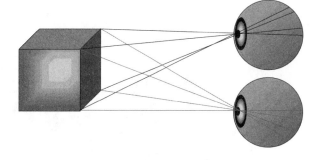

Figure 2-6 *Our two eyes capture slightly different images of the same thing.*

Conventional cameras have moved on from using pinholes. Because pinholes only let a small amount of light into the camera, as you saw when looking at the image on the screen, they're not very good for getting sufficient light to the film or digital sensor. Lenses allow more light into the camera, as we can see in Fig. 2-5; however, they can only bring part of an image into focus at any one time. This effect can be used in photography to help create a feeling of distance. For example, the near objects are sharply in focus while distant objects are blurry. However, in holography, we aren't using lenses to "focus" an image, and the amount of depth that can be crisply captured depends on the "coherence length" of our laser.

One of the things we can see with our pinhole camera is the illusion of parallax. As we move, objects in our field of view, depending on their distance from us, appear to move relative to each other as we change our point of view from side to side. Let's illustrate with an example. From where I am sitting at the moment in my office, I can see beautiful Welsh mountains in the distance and my monitor in front of me. The mountains are much further away than my monitor. As I move my head to the right, the mountains appear to move "right" with me while my monitor appears to move left. The converse is true: When I move my head left; the monitor darts right in the opposite direction while the mountains move left with me. Here is a crucial point: Changing your position, things appear to move horizontally at different speeds relative to their distance from the observer.

Our eyes are quite close together. This distance is referred to by some people as the *interocular separation,* and when we build a stereo camera, we try to mimic this by placing the lenses at a similar distance apart. However, imagine an alien with eyes positioned much further apart. The parallax effect would be that much greater due to the different viewpoints.

Moving our head from left to right gives us *horizontal parallax.* A similar effect is also achieved by moving our head up and down: *vertical parallax.*

We can see in Fig. 2-6 how, when focusing on an object, our two eyes capture different images of the same object. We can see this effect by closing one eye, then opening the other eye while closing the one that is open rapidly. By repeating this in quick succession, we see the image "jump about" as the two eyes capture two different images. In binocular vision, our brain takes the information from both and fuses it together to make a 3D image.

Viewing Stereo Pairs

We've discussed how we see in three dimensions, so let's try and view a stereo pair of images. We can then see what makes this different from viewing a hologram once we've created some 3D images.

Figure 2-7   *A stereo pair of images.*

You can buy a special lens, a stereo pair viewer, which will help immensely in directing the focus of your eye to each individual image. However, here is a simple equipment-free straight viewing stereo method that you can use to view the stereo pair in Fig. 2-7.

While you look at the page, the idea is to relax your eyes so that they perceive the two images at once as a single image. Stare at the book, concentrating each eye on its own respective left or right image. Now move the book closer toward you. Rather than adjusting your vision to focus on the book's page, try and look "through" the book. Eventually you will see the images double up, and a third image form in between them, which is the composite image in three dimensions. Slowly try and move the book away from you without losing focus—lock your gaze! With a bit of patience you can master the art of "free viewing" 3D images. If you can't, however, don't lose hope! Buy a viewer, and you will find it much easier. Figure 2-8 shows the divergent free-viewing method.

The beauty of holography is that it requires no special viewing equipment (only a laser for some types of holograms), and you don't need to master

Normal Reading / Photograph Viewing

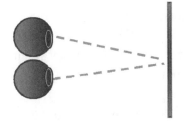

The "Divergence / Magic Eye" Method

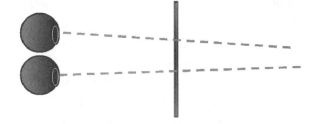

Figure 2-8   *The divergent free-viewing method.*

any special knack to perceive the 3D effect. It's worth understanding the different types of 3D imaging technology because they provide a contrast to holography and gives us the perspective (if you'll forgive the pun) to understand how holography is different and what makes it so special as a technology.

Chapter 3

Basic Practical Optics

Optics are the key tools of the holographer, and they will be indispensable in helping you to move, bend, and shape light to produce glorious three-dimensional (3D) holograms. In this chapter, we explain all of the basics that you need to know about working with optics. How different elements function is the first half of the chapter, and the second half focuses (if you'll forgive the pun) on how to mount these optics and use them in a practical context.

Mirrors

When setting up for making holograms, it is useful to be able to bounce the light around the sandbox or optical bench. The easiest way to do this is with some front surface mirrors. They are called front surface mirrors because, unlike ordinary mirrors—for example, the sort you might find in the bathroom—front surface mirrors have their "silvered coating" applied to the front rather than the rear. There is a reason why we use front surface mirrors rather than ordinary household mirrors. Because a household mirror has the coated surface behind a piece of glass, the laser beam must first pass through the glass before reaching the mirror. It is then reflected and must pass back through the glass before continuing on its journey. Traveling through all this glass degrades the quality of the beam and also can cause secondary beams to be formed. You can see this illustrated in Fig. 3-1.

Try and obtain some different sizes of mirror. Small pieces of mirror are OK if you want to bounce ordinary straight laser beams around; however, if you are trying to reflect a beam that has been spread out by a diverging lens, then you will want to have a few larger pieces of mirror on hand.

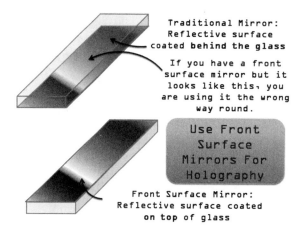

Traditional Mirror: Reflective surface coated behind the glass

If you have a front surface mirror but it looks like this, you are using it the wrong way round.

Use Front Surface Mirrors For Holography

Front Surface Mirror: Reflective surface coated on top of glass

Figure 3-1 *How a front surface mirror and household mirror differ.*

Cleaning Front Surface Mirrors

Because the coating of a front surface mirror is exposed, rather than being protected behind a layer of glass, it is vulnerable to being scratched and covered in dirt. Try and handle them on the edges so that fingerprints do not accumulate on their surface. If for any reason you do need to clean them, the best thing to use is alcohol, which will help remove grease and not make the surface of the mirror moist. Remember, this is a silvered coating, so moisture may cause it to tarnish. Keep it dry at all times. You would be sensible to store your front surface mirrors between layers of tissue to protect them in storage. If you have some silica gel on hand, often found in boxes of new shoes where it acts as a desiccant, store this in the box with your mirrors because it will help absorb moisture and stop them from tarnishing.

You Will Need

- Front surface mirror
- Sheets of newspaper

Tools

- Glass-cutting tool
- Steel ruler

The glass-cutting tool consists of a strong wheel mounted to a metal handle. You can see the tool in Fig. 3-2.

You will need a number of mirrors for your holography setup, and the cheapest way that you can source these is to find a big piece of front surface mirror and cut your own. This technique can also be applied to beam splitters. In their simplest form, they are a piece of regular glass; however, it is also possible to buy pieces of fancy beam splitter that contain silvered surfaces that will split a beam in an exact ratio.

To cut mirror successfully, you will need to lay the mirror on some sheets of newspaper. Hold the mirror firmly and using the steel ruler to make a straight line, run the glass-cutting tool over the surface of the mirror making a groove on the surface of the glass. Once you have done this, you can take the scored mirror, place it over something hard (like the handle of your glass-cutting tool), and apply firm pressure on either side. Your piece of glass will snap into two pieces along the line you have scored (Figs. 3-3 and 3-4).

When pressing the front surface mirror, be careful not to touch the side that is mirrored (Fig. 3-5). Apply pressure to the "backing" side because the grease and moisture from your fingers will damage the mirrored surface.

It's important to understand that if we were to draw a line perpendicular to the mirror's surface

Figure 3-2   *A glass-cutting tool.*

Figure 3-3   *Scoring the line on the glass.*

Figure 3-4 *Applying firm pressure to either side of the scored line.*

(called the normal), the angle of reflection will always be equal to the angle of incidence. A diagram makes this a little easier to understand; see Fig. 3-6.

In addition to your straightforward common or garden planar mirrors, it's also important to note that sometimes we need something a little bit unusual. Think about the curved mirrors you might see in a fun house at a fairground and how they distort and bend light.

One thing worth keeping an eye out for is a concave mirror. This is useful in conjunction with a lens when we want to form a collimated beam of light, which is helpful in some advanced

Figure 3-5 *Do not touch the mirrored side of front surface mirrors.*

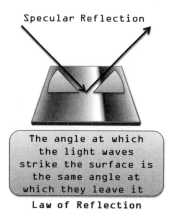

Figure 3-6 *How light is reflected from planar mirrors.*

holography setups. In particular, look for a concave mirror that is "part of a sphere," also called a spherical mirror. You can see an example in Fig. 3-7 and an example of what a setup would look like to produce collimated light in Fig. 3-8.

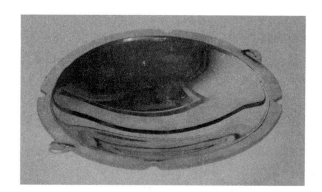

Figure 3-7 *Concave spherical mirror.*

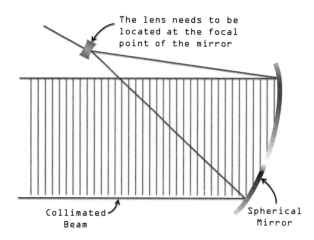

Figure 3-8 *Producing a collimated beam of light using a lens and spherical mirror.*

We also will employ a convex mirror in the advanced holography projects chapter, where we will use it to spread out a beam to produce a 360° hologram.

Beam Splitters

We often have a situation in holography setups where we want to create two laser beams, an object beam and a reference beam, from a single laser beam. The way to do this is with a beam splitter. Figure 3-9 shows an example of a beam splitter mounted to a simple plastic optical mount.

If you are just starting out, you can make perfectly acceptable beam splitters from an ordinary piece of glass. Try and source a thick piece of glass. You will find that when shining a laser through the glass, secondary reflections are formed. You'll need to use some black electrician's tape to mask them off (Fig. 3-10).

If you have deep pockets, you can also buy a professional "adjustable" beam splitter. This

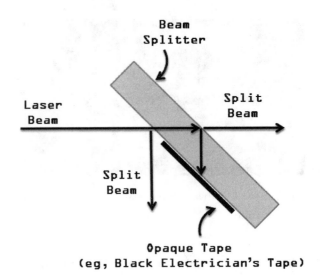

Figure 3-10 *Masking off secondary reflections from a beam splitter.*

consists of a "wheel" of beam splitter material, with a silvered coating applied whose density varies depending on the angular position. By turning the wheel and aiming a laser beam through different sections of the wheel, it is possible to split the beam into two parts of varying proportional brightnesses that change depending on the angle you "dial in" with the beam splitter.

Lenses

A helium-neon laser produces a straight line beam, which will produce a spot when shone on a surface. In the process of making holograms we want to spread this beam out in order to illuminate the whole of the object and the plate.

Diode lasers get around this problem for simple holography. If you can find a diode laser with a "removable collimating lens," then you can unscrew it to produce an oval beam pattern directly from the diode laser. The job of the collimating lens is to take the spread-out beam from the diode laser and concentrate it to a "beam."

It's clear, therefore, that we need to get to know the different types of lenses. Lenses are either

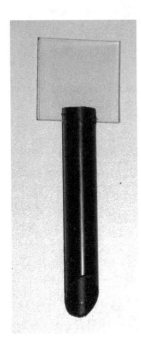

Figure 3-9 *A beam splitter fashioned from a piece of glass and piece of plastic pipe for support.*

Figure 3-11 *Convex lenses focus the beam to a point, whilst concave lenses spread the beam.*

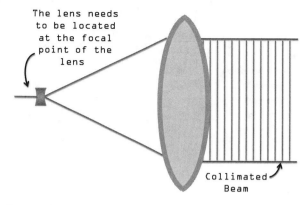

Figure 3-12 *A collimated beam from a small concave lens and a large convex lens.*

"divergent" or "negative" lenses, or they are "convergent" or "positive" lenses. The two simple types of lenses that you are most likely to encounter are "concave" lenses and "convex" lenses (Fig. 3-11). Convex lenses are used to focus a wide beam to a point, and concave lenses are used to take a thin beam and spread it out. Although this is a simplification, it's useful to have this basic view of lenses to start with, before getting into more complicated arrangements of lenses.

To complement this basic understanding, it's also important to understand that "focal length" is a measure of the "bending power" of the lens. The shorter the focal length of your lens, the more strongly it will spread out or converge in your laser beam.

We can use compounds of lenses to achieve different effects, and these are explained as necessary in the holography projects section. As one example, we looked at how we might use a mirror to produce a collimated beam; however, equally, we could achieve the same effect using a small concave and large convex lens (Fig. 3-12). One of the things that will become apparent in your holography experiments is that there are often separate routes to achieving the same thing.

Shutter Cards

Unless you are going to have a sophisticated holography setup with an automated timed beam blocker (see Chap. 13 for details on how to make one), you will need some way of blocking the laser beam to time your holography exposures. At the very simplest end of things, you can simply use your hand to block the beam, moving it out of the way to make an exposure. However, chances are you'll want your hands free to do other things. Keep a supply of thick black card because it will absorb the beam and prevent unwanted reflections that could lead to exposure of your holographic plate.

White Card

Sometimes we might need a sheet of white card to use as a screen onto which we can project laser light (see the science fair projects). Also, before you crack open your holographic film, or plates, it is useful to have a piece of white card to use in its place when aiming laser beams.

Spatial Filters

Despite the holographer's best efforts to keep lenses and optics clean, there is always the

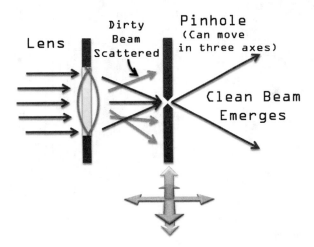

Figure 3-13 *Diagram of spatial filter.*

propensity for them to get dirty and grimy, which will ultimately reduce the quality of the beam produced. Spatial filters are the answer to all your dirty beam woes. They are particularly awkward to make yourself, so if you are going to use them, you will do well to buy them ready made.

The lens in the spatial filter concentrates the beam and focuses it to a single point—a pinhole is placed exactly at this point. Imperfections in the lens, dirt and grime on the lens, or a "dirty" beam that results from passing through other dirty optics will cause the beam to scatter as it passes through the lens. Because only the "clean" beam can pass the pinhole, this scattered light that would cause noise in a final hologram does not pass through the focused pinhole and so is "removed" from the emergent beam. This process is illustrated in Fig. 3-13.

Mounting Optics

The optical elements provide the utility of bouncing the laser beam about and shaping it to allow us to control its output in order to make holograms; however, we need to devise a method to rigidly mount the optics to hold them still while we make exposures. The smallest movement of an

optical element can result in disaster for our holograms, and so it is immensely important that we control vibration wherever possible.

A number of different avenues are open to the hobbyist holographer when it comes to mounting optics. Let's start by looking at what the professionals use to draw inspiration, and but then we'll work our way down to practical solutions that are more affordable and represent good value for the hobby holographer.

The professionals use an optics bench, a heavy steel and concrete construction that provides holes to affix optical elements. If you have access to one of these at a local college or university, it could present a good option, especially if the school has all of the fixtures and fittings to use with this bench. Sometimes a quiet word with a friendly technician will get you access to an optics workshop after hours.

Next down from a full flat optics bench is a linear optics bench. You can see a couple of examples of how a linear optics bench could be set up to make simple reflection holograms in Figs. 3-14 and 3-15.

As you can see, the linear optics bench takes the form of an extruded length (usually of aluminium)

Figure 3-14 *Optics bench in use for making simple reflection hologram.*

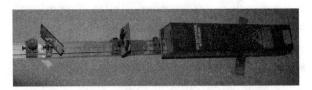

Figure 3-15 *Optics bench in use for making simple reflection hologram.*

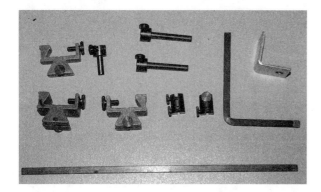

Figure 3-16 *A range of different accessories for use with an optics bench.*

Figure 3-17 *A range of optics mounted in 35-mm slide mounts.*

to which you can bolt "saddles." These saddles have a number of screws that can be used to affix them rigidly to the bench, and then provision is made to attach different fixtures. For example, saddles usually have a slot cut in the top, which can be used to attach flat objects, and lenses mounted on flat pieces of plastic, mirrors, beam splitters, or rods in turn can be used to support other optics. You will generally find an array of hardware supplied with your optics bench: flat pieces of metal bar, connecting pieces to join bars, support poles, and so on (you can see quite a range in Fig. 3-16), all of which, with some creativity and ingenuity, can be used to support the equipment required to make holograms.

For one of their optics courses, the Open University developed a fantastically versatile system, from which the amateur holographer can learn a lot. All of the optics are mounted in pieces of plastic the size of 35-mm slides (Fig. 3-17). If you find a specialist photographic dealer, it is often possible to buy a range of slides, with apertures shaped for different size circular holes, which can then be used to glue a selection of lenses into. Although film-based slide shows are less popular now because we are in the digital age, a quick look on the Internet or a handful of phone calls to a number of small businesses listed in a photography magazine, and you should strike gold! Once you've mounted your lenses on these little sheets of plastic, you have a versatile system that can be

used with a range of holographic setups: You can mount them in plastic pipe, use them with an optics bench—the list is endless—and because you handle the slide, not the lens, you keep the optics free of fingerprints or imperfections.

There are some limitations to using a linear bench. Although for a professional piece of kit, they are within the budget of a hobbyist holographer, they do not permit much deviation from a linear format. So beyond making simple reflection holograms, you're going to have to get a bit creative, and some of the more complicated setups are ruled out entirely. However; they do have their applications, and if you are only going to undertake relatively simple experiments, they should not be wholly discounted.

Now we reach the realm of true hobbyist holography and things you can build yourself. Without doubt, one of the most flexible routes into holography is to employ optical elements that can be used with a sandbox. There are constructional details on building your own sandbox in the chapter "Your Holography Workshop," and building optical mounts to use in a sandbox is really as simple as sticking something (anything!) into a layer of sand.

If all this sounds like too much trouble, you can buy some off-the-shelf mounts that will work well with a holography sandbox. My own favorites come in the form of a kit from Industrial Fiber Optics (www.i-fiberoptics.com). Their system uses punched and bent metal supports. These take the form of a "U," so either side of the optic is supported while light can pass through the middle. These supports are really nice to use; a strip of magnetic plastic with a self-adhesive back is supplied, which can then be affixed to the back of mirrors and to the side of beam splitters (Figs. 3-18 and 3-19).

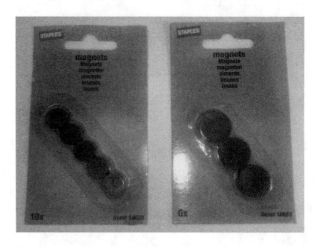

Figure 3-20   *Small magnets for temporarily affixing optics.*

Magnets (of the kind shown in Fig. 3-20) are immensely handy for making things stick together rigidly when you need them to, and they are easy to disassemble when you don't. It makes sense to pick up a couple of packs of different size magnets because you can easily hot-glue them to things that you want to join together temporarily.

Another great innovation is magic magnet magnetic tape, which makes it exceptionally easy to make almost any flat surface you want magnetic. It isn't quite as strong as the small magnets that you need to glue on, but for light thin mirrors or piece of glass to use as beam splitters, magic magnets (shown in Fig. 3-21) should be more than adequate.

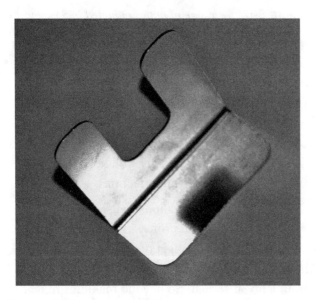

Figure 3-18   *Industrial Fiber Optics metal optic mount.*

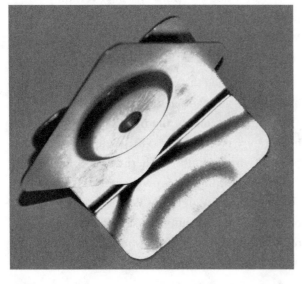

Figure 3-19   *Lens mounted in Industrial Fiber Optics metal mount.*

Figure 3-21   *Magic magnet tape.*

Figure 3-22 *Affixing magic magnet tape to the back of a small mirror.*

It's simply a matter of peeling of a length of the adhesive magnet tape, which is pre-scored so you can easily rip it to length, and sticking this onto the back of the thing you want to magnetize (Fig. 3-22).

A bit of hot melt glue and a wooden stake will produce the simplest of mirror mounts, as you can see in Fig. 3-23.

It's also worth sticking some small clips to wooden stakes to make versatile mounts that can hold a variety of flat objects, for example mirrors, as depicted in Fig. 3-24, or for example, optics mounted in 35-mm slide housings as described earlier.

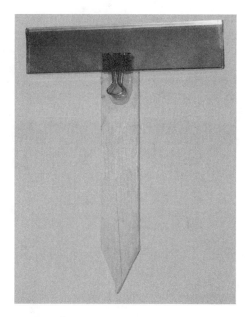

Figure 3-24 *A versatile clip mount.*

If you want to get a bit more advanced; the polyvinyl chloride (PVC) pipe commonly used for electrical conduit or in plumbing for the overflow pipes of sinks and toilets is incredibly useful and versatile, and it comes with a whole range of off-the-shelf fixtures and fittings (eg, right-angle bends, tees, pipe wall mounts), all of which can be cannibalized to make improvised (but nonetheless very durable) optical mounts. When you go to your plumbing supply outlet, be sure to also pick up some "pipe weld" adhesive (shown in Fig. 3-25),

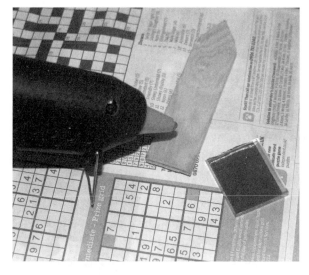

Figure 3-23 *Making a simple mirror mount from a wooden stake and hot melt glue.*

Figure 3-25 *Pipe weld adhesive.*

Figure 3-26   *Cutting plastic pipe at a 45° angle on a circular saw.*

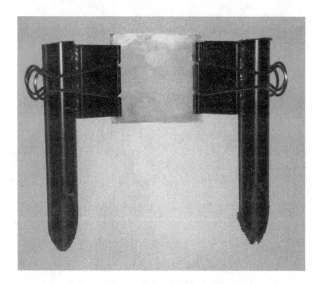

Figure 3-27   *A holographic plate support.*

which is used to join plastic plumbing fixings, because it will help you make rigid durable mounts.

When you are ready to make optical mounts, the easiest way to do it is to cut several lengths of pipe—say 10 to 12 in/200 to 300 mm—and then, using a circular saw as shown in Fig. 3-26 (for speed, but other saws will do), cut a diagonal cut through the center of the pipe, leaving you with two equal size pieces. It makes good sense to prepare loads of these at the same time. They don't take long to make, and having a handy box with some spares in is useful when you need to fabricate something unusual.

Then, when it comes to mounting optics, it is a case of cutting a slot in the top, for example when you want to mount a mirror or beam splitter, an example can be found in Fig. 3-9. Or for circular optics, a tee or right-angled bend can usually be cut away and fashioned such that a circular lens can be supported.

Do not underestimate the utility of simple clips available from stationery stores everywhere. A healthy supply of bulldog and spring-back clips will find a multitude of applications in securing mirrors, beam splitters, holding holographic plates, and securing flimsy holographic film between two glass sheets. The list goes on. For the minor

investment in a basket of office supplies, it is worth having a look around your local store for improvised holographic accessories. Keep things on hand, and you will be grateful for having the foresight to buy them when it comes to securing your optics. If you want to see an example of how to use these, look no further than Fig. 3-27, which shows a simple holographic plate holder fashioned from two pieces of PVC pipe and some spring-back clips.

Again, hot melt glue is fantastically versatile, in that not only can it secure just about anything to anything, but it also has sufficient play, allowing you to build up surfaces so your optics can be beautifully aligned.

One other useful system, which will work well with sandboxes, is to use hot melt glue to attach a standard camera "tripod mounting nut" onto your optical components. You can get fixtures and fittings, such as those illustrated in Fig. 3-28, which you can customize to your purpose.

A dab of hot melt glue secures the optic in place, as you can see in Fig. 3-29, making something that can be attached to any accessory with a camera mounting screw. There are such a plethora of tripods and mounting devices available

Figure 3-28   *Tripod mounting screw.*

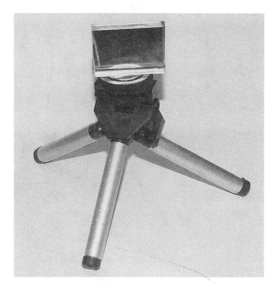

Figure 3-30   *Mirror mounted on mini tripod.*

Figure 3-29   *Hot gluing a camera mounting screw onto the back of a mirror.*

some cases flexible tripods, with legs that can be bent to conform to the shape you require, as in Fig. 3-31. These are sometimes useful if you need to achieve a special effect, such as mounting a mirror over an object.

that these can come in immensely handy if you need to achieve an awkward mounting position or if you need to position a mirror "off board" if you find your sandbox is too small (although be aware that anything not mounted in your sandbox will be extra susceptible to vibration!).

Then buy some cheap "mini tripods" from a photo shop; these come as small tripods that can be folded into shape, as shown in Fig. 3-30, or in

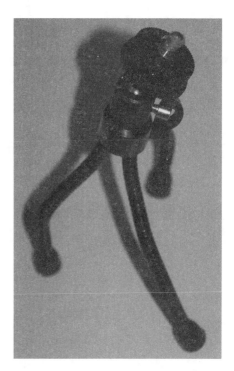

Figure 3-31   *A flexible tripod useful for achieving awkward angles.*

Figure 3-32   *Foam board can be used to quickly fabricate optical mounts.*

Finally, a rigid material that is useful to have at hand is foam board. It is incredibly versatile and can be easily cut with a scalpel and made into almost any rigid shape. Because the board has some "thickness," it is easy to glue together with hot melt glue to form strong 90° joints. It's often worth prototyping something in paper first of all to see how it will fit together; then you can transfer your design to foam board. The support shown in Fig. 3-32, used to hold a holographic plate and model at a vertical incline, was made from foam board quickly and cheaply.

Chapter 4

Light and Lasers

If you've read about the history of holography, you'll realize that Dennis Gabor's early holograms were limited by the lack of a decent light source. There's something very special about laser light that makes it particularly well suited to holography. We'll see a little more about this in Chap. 5, which explains how holograms work, but for now, let's take a look at light and lasers.

Understanding Light and Why Lasers Are Different

When we look at the light coming from an incandescent bulb, for example in Fig. 4-1, we see that it is of a multitude of different wavelengths, none of which are in phase.

Whereas if we look at light from a laser, as illustrated in Fig. 4-2, one thing strikes us: All of

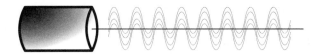

Figure 4-2 *Light from a laser.*

the light coming from the laser is of the same wavelength. That is, there is the same distance between the peaks and the troughs of each different wave. Furthermore, you'll see that these peaks and troughs coincide with each other and move in step. This is known as *coherence*.

Light is part of the electromagnetic spectrum, and light, in common with all other electromagnetic waves, is composed of an electric and magnetic field that propagates as a pair of sinusoidal waves at right angles to each other. We can see this in Fig. 4-3. In this book, we use convention and illustrate light as a single sinusoidal wave, or, in some cases, we indicate the wavefronts with a series of parallel lines.

By adding energy to a laser, we can cause it to produce electromagnetic radiation of a certain

Figure 4-1 *Light from an incandescent bulb is of all different wavelengths.*

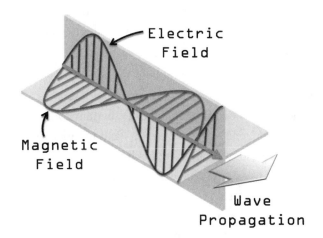

Figure 4-3 *Electromagnetic waves.*

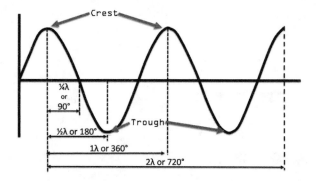

Figure 4-4   *Wavelengths explained.*

wavelength. The laser itself isn't producing any energy—that would be impossible. However, the laser as a device can take out electrical energy, and transform it into electromagnetic radiation of a certain wavelength—visible light for our holograms.

Light waves travel at—guess what—the speed of light, but if we were to zoom in and look impossibly close, we could see that our wave fluctuates up and down. To give us a measurement, we can look at the *wavelength,* which is to say, the distance between any two successive points that repeat in the cycle. Figure 4-4 should add a little clarity to the matter. Furthermore, if we were to pick a point in space and measure the number of "peaks" or "troughs" passing past this point (and you'd have to be quick because light travels at 299,792,458 km per second or 186,282 miles per second, we would see that by counting the peaks we could determine a *frequency,* or how many peaks pass that point in a unit of time—let's say a second). We can see, therefore, that if light propagates at a constant velocity, frequency and wavelength are related.

We will come across these terms again. As when we are discussing lasers, we know that lasers generate monochromatic light of a certain color—a single frequency or wavelength—and in order to be more precise, we can term this light "coherent," which is to say they have a "constant relative phase," or in simpler, less scientific terms, as the light exits the laser at the speed of light, the peaks are all synchronized, and the troughs are all synchronized.

The Laser Explained

Take a look at Fig. 4-5. We're going to use this as a simple diagram to help you understand the laser. Our laser consists of a resonator, which is a cavity that contains a *gain medium*. In the case of our ruby red lasers (the type that Theodore Maiman made), the gain medium is ruby; in our helium-neon gas lasers, the gain medium is a mixture of, well, helium and neon gases, and in the case of our semiconductor diode lasers, the gain medium is a semiconductor.

At each end of this gain medium is a mirror; on one end the mirror is fully silvered, so all the light is reflected; at the other end the mirror is partially silvered. What this means, in effect, is that most of the light is bounced to and from between the two mirrors, but a small portion of it escapes, and this is what we see as a "laser beam." Now we need some way to stimulate this emission, a "pump," if you will. In the case of Maiman's original laser, the pump came in the form of a flash, a bit like the kind of flash you'd find in your camera. However, with a helium-neon laser, we create this pumping effect by exciting the gas with a high voltage, producing the characteristic red discharge that gives neon signs and our helium-neon laser its warm light. In our semiconductor laser diodes, the pumping effect is created by a flow of electrons and holes through a semiconductor material. The mechanisms are subtly different, but the net effect is the same: the production of bright laser light.

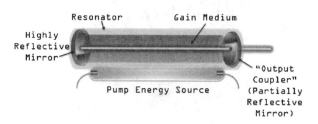

Figure 4-5   *The laser explained.*

Skip ahead to Fig. 4-14 if you need reminding what an atom looks like. (In fairness to modern science, this is a classical model that acts as an aid to understanding; our modern view of science is that electrons do not have "discrete orbits," but visualizing it in this way is a simplification that helps us understand the processes that underpin the laser.) So to recap, for those of you whose chemistry is rusty, atoms consist of a *nucleus* that contains the protons and neutrons, and an orbiting cloud of electrons. We can see that the electrons circle at a number of different orbits or energy levels. Atoms can be at different energy levels; the basic pattern of things is that they are at a "ground state energy level"; however, if we apply a lot of energy to them, they can reach an "excited" energy level.

So we can excite atoms by adding energy, which could be in the form of heat, light, or electricity. By exciting atoms, we help their electrons to move from lower energy level orbits to higher energy level orbits. Thus we are "pushing" the electrons away from the nucleus. We can see this in Fig. 4-6.

However, this situation cannot remain forever; once in a high energy level orbit, the electron still isn't satisfied, and its natural tendency is to return to the ground state. We can see this clearly in Fig. 4-7.

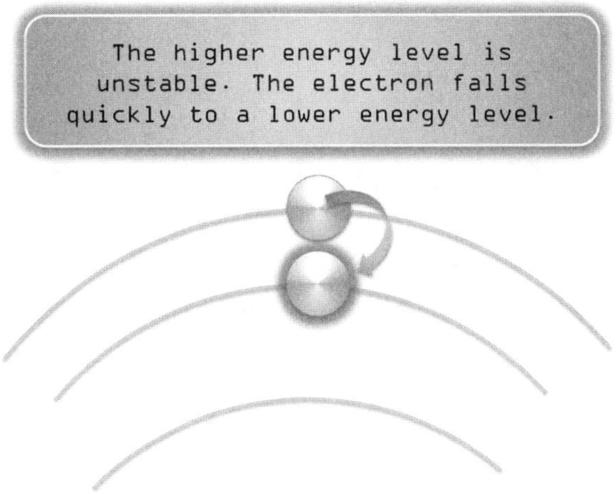

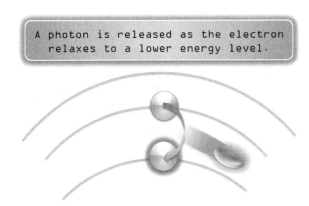

Figure 4-7 *The electron's natural tendency is to fall to a lower energy level.*

As the electron drops an energy level, it releases a photon—a little bit of light. There are lots of mechanisms for photon emission. When we see the filament of an old-style incandescent light glow, this is the product of atoms, which are excited by the thermal energy of wire flowing through the filament. As these atoms are excited, their electrons change energy levels, and in the process of doing so, they emit photons (Fig. 4-8).

This process is taking place all over the gain medium, and it can so happen that a released photon collides with an electron at an elevated energy level, which then relaxes to a lower energy level, releasing two more photons in the process.

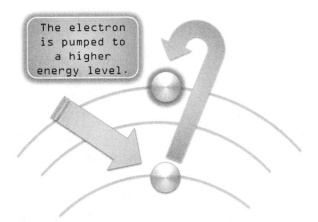

Figure 4-6 *The electron jumps to a higher energy level.*

Figure 4-8 *Relaxing to a lower energy level causes a photon to be emitted.*

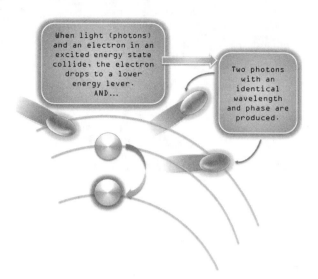

Figure 4-9 *Photons collide with electrons to release more photons.*

This is the process of *amplification,* which quite literally puts the "A" in laser. We can see this in Fig. 4-9.

As we saw earlier, mirrors at either end of the laser help bounce light back and forth (Fig. 4-10). This keeps the photons stimulating the process and helps sustain the lasing. One mirror at one end is a completely silvered mirror. However, at the opposite end of the laser is a partially silvered mirror, which allows a little bit of light to escape, and this is what we see as the laser beam.

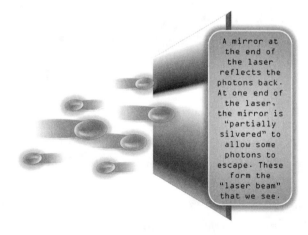

Figure 4-10 *The mirrors at the end of the laser bounce the photons back and forth.*

Figure 4-11 *Helium-neon laser opened up.*

Gas Lasers

It helps to have a look inside a helium-neon laser to understand what is going on. Warning: There are high voltages to be found inside a helium-neon laser. I've taken one apart to show you what is inside so you don't have to. You can see it all in Fig. 4-11.

We can see at one end the glass tube that contains the helium-neon gas and mirrors at each end while at the other end of the enclosure is the high-voltage power supply, which provides the potential differential required to pump the laser tube.

Semiconductor Lasers

One of the things that has really helped make holography cost effective and accessible to the hobbyist is the introduction of reasonably priced semiconductor lasers, such as the laser pointer shown in Fig. 4-12.

The path to getting a semiconductor laser that would work continuously and at room temperature was a long and arduous one, briefly documented in Chap. 1 on the history of holography. However, since their first practical demonstration they have come down in price to the point where they are ubiquitous. They are found in many modern electrical devices, from the bar code scanner to the blue-ray video player. But how do they work? The principles behind their operation are not dissimilar to light-emitting diodes, but how does an LED work, and how do laser diodes differ?

Figure 4-12 *Laser pen employing semiconductor diode laser.*

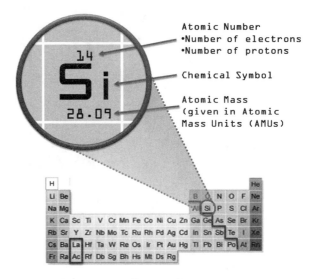

Figure 4-13 *Silicon and its place in the periodic table.*

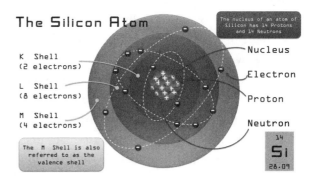

Figure 4-14 *The atomic structure of silicon.*

Bear in mind that laser diodes tend to use semiconductors based on sophisticated semiconductors based in turn on alloys of such chemicals as gallium, arsenic, phosphor, indium, and aluminium. However, because we're only just starting out to understand the chemistry of semiconductors, let's pretend that our laser diodes are made of silicon. It will make the chemistry much easier to understand. The LEDs that you see are based on silicon, and they're not a million miles away from what is going on in a laser diode.

The story all starts with the element silicon. Silicon is a semiconductor; which is to say a material that has neither the electrical properties of a conductor or an insulator, having an electrical conductivity somewhere in between. Its atomic number is 14, which tells us the number of protons and electrons that silicon ordinarily possesses (Fig. 4-13).

If we look in a little more detail at silicon, we can see that the outer shell of electrons, commonly known as the *valence shell,* ordinarily contains four electrons (as shown in Fig. 4-14). We sometimes also refer to this shell as the *M shell.* The valence shell, or M shell, of any chemical

element is important because it gives us clues as to how those atoms will join together and how reactive the chemical is. In all chemical reactions, chemicals strive to "complete their outer shells." In this case, silicon strives to fill its "M shell" with eight electrons. (The noble gases in group eight all have full outer shells of electrons, and so they are very unreactive.)

The way that silicon can do this is by linking with other silicon atoms, sharing electrons in the process, and thus fill the outer shell. Silicon is a crystalline solid, which means the individual atoms of silicon align themselves in a regular pattern to form a matrix of silicon atoms, which because of their configuration we call a *crystalline structure* (Fig. 4-15).

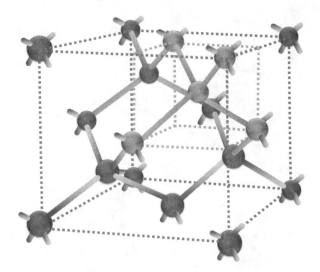

Figure 4-15   *The crystalline structure of silicon.*

Ordinary silicon is not particularly conductive. To control a flow of electrons and make silicon useful, we need to add some impurities. These impurities effectively add extra electrons, or take electrons away from parts of the silicon, creating inequalities and imbalance. We can then use these inequalities to control the flow of electrons through the material, and by doing so make useful things like transistors, integrated circuits, diodes, and, importantly for us holographers, *laser diodes*.

N- and P-Type Silicon

So one type of silicon we can make (as you can see in Fig. 4-16) is known as "N-type" silicon. We call it N-type silicon because it carries an overall negative charge. It has gained some extra electrons along the way. The reason for gaining extra electrons is because it has been "doped" with tiny amounts of phosphorus. Phosphorus carries five electrons in its outer shell.

Another type of silicon we can make is known as "P-type" silicon (see Fig. 4-17). P-type silicon is so called because it carries an overall positive charge as the result of missing electrons. These

missing electrons come from the fact that we dope the silicon with an additional little bit of boron, a chemical that only has three electrons in its valence shell. As a result of this, where boron is found in the crystalline structure, there is a missing electron, or electron hole, in the region of the boron atom, and so because some "negative charge is missing," the region carries a positive charge.

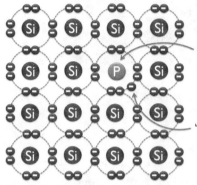

N-Type Silicon has a little extra Phosphorus added, which gives the resulting material "extra" spare electrons. These can move easily throughout the crystal.

"SPARE ELECTRON"

Figure 4-16   *N-type silicon.*

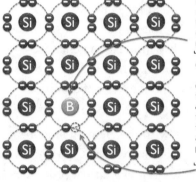

P-Type Silicon has a little bit of Boron added, which gives the resulting material "holes," or missing electrons. This hole can move easily throughout the crystal because an electron can fall easily into the hole, leaving a new hole in its place.

"Electron Hole"

Figure 4-17   *P-type silicon.*

So we then create a sandwich, so to speak, of these different types of semiconductor. Think of the sandwich in terms of different color bread. This isn't just your plain old sandwich made with one slice of white bread and one slice of brown, but it's a super sandwich with several layers making a striped pattern of white, brown, white, brown.

A little voltage is needed to start the holes and electrons flowing. Where the N- and P-type silicon meet, we have a region called the *space charge region,* which you can think of as a no-man's-land between the N and P types of silicon.

When the positively charged holes do eventually move to combine with the negatively charged electrons, energy is released in the form of photons—or, if you prefer, light. Now in a straightforward PN junction, you would release this light, and you'd have an LED. However, in a laser diode, we have something special, a channel formed in the material of the laser diode, with a mirror built into each end (Fig. 4-18). As a result, we get the same amplifying effect that takes place in a helium-neon laser, or ruby laser, where light bounces back and forth between these two mirrors, being amplified as it passes by other electrons that

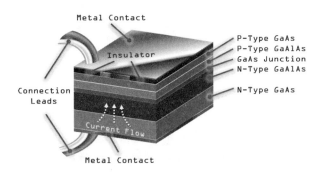

Figure 4-18 *Heterojunction laser diode.*

are energized. The structure of the laser diode is such that when it emerges from the partially silvered mirror, it diffracts, which is another way of saying it spreads out to form the characteristic cone shape that you will see from a laser diode with its collimating lens removed. The cheap laser pointers that you see employ an additional lens to help shape or collimate this beam into the thin red line that we know as a laser beam.

Sourcing Your Laser

Sourcing lasers used to be as rare as finding cheap lawyers. However, lasers today are relatively cheap. For some of the simpler projects in this book, a simple diode laser with the lens removed will suffice. For some of the more complex projects, a helium-neon laser would be a far better bet. The online marketplace has helped make sourcing secondhand lasers and finding specialist components easier than ever. In the back of this book is a supplier's index that contains a plethora of different sources, or you can check out the Web site for current links. If you're interested in building lasers, there is some additional content at www.holographyprojectsfortheevilgenius.com that will guide you in building your own laser power supplies.

Chapter 5

How Holography Works

We've looked at some simple optics, we know how lasers work, and we've learned about some of the background of holography. We're now getting to the point where we're ready to make holograms, but how do they actually work? If you want to dive ahead, you could tackle some of the simple holography projects first of all and then revisit this one. However, if you want a firm underpinning of theory, then wade through this one first of all, and your holograms will only benefit from this newfound knowledge.

The Properties of Holograms

Holograms record information quite unlike any other form of imaging. When we look to a history of imaging using light-sensitive emulsions, we go back to the days of the black-and-white photography. Black-and-white photographs record information only about the "intensity" of the light falling on a light-sensitive surface. As a focused image, the information is only recorded in two dimensions and captures a static scene. With the advent of color photography, additional information was captured pertaining to the wavelength of the light. The wavelength of light determines its color, and so by using a coating system containing different emulsions—each sensitive to different wavelengths of light—it is possible to capture information about the colors in the scene. Again, this is a focused image of a three-dimensional (3D) scene, captured on a two-dimensional (2D) plane.

So what makes holography different from conventional methods of recording images? For a start, we are recording an "unfocused" image. Rather than recording a static image from a single position, we are recording a "field" of light reflected or transmitted from or through the image. What arrives at each point in the emulsion is the "sum" of all the light that scene is reflecting to that point. So why don't we just end up with a messy blur? What makes holograms different is that we are recording the "phase" information, which is to say, because the light coming from the laser is "coherent," that is, all of the waves leave the laser "in phase." By the time they have reflected off the object in question and reached the emulsion, the different parts of the image will receive a multitude of light waves all at different phase points. Because we can add and subtract wavelengths; as we will see later, we can combine the "object" beam, that is to say the light reflected from the object, with a "reference beam," to produce a recording that captures both the intensity of the light reflected from the object and the phase when it hits the emulsion. This is information that is lost in conventional photography. Figure 5-1 provides a summary.

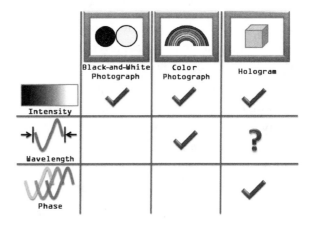

Figure 5-1 *What makes holography different from other forms of imaging.*

If the last paragraph made no sense, then read on. Then skip back and reread it! All will become, well, coherent!

It's beyond the scope of this book to explain color holography in any detail, but it is worth quickly mentioning that with sophisticated optics setups and some very expensive lasers, we can use multiple lasers that emit different wavelengths of light, combined, to produce a hologram that will capture wavelength information and provide some color data about the object in question.

Because we have captured all of this additional phase information, we can reconstruct the field of light produced by the object, by replaying a hologram with a reference beam, reproducing a 3D scene. But what else is unusual about holograms?

As we will see in the chapter on science fair holography projects, holograms also have the unusual property that if we make a hologram of an optical element—for example, say we make a hologram of a magnifying glass—the hologram will perform in a similar manner to the original optic.

Hologram, as we have seen, means "whole image," but unlike a photograph, which you could rip in half and separate into two different images, holograms have the unique property that every point on the emulsion contains information about the whole film. This means that we can cut our hologram into pieces, and each piece will contain image about the whole image—albeit we might have to adjust our viewpoint to see the scene.

Similarly, a scratch or imperfection in a picture means that information is lost forever, and only by using a photo-editing program can we reconstruct the lost data. And even then, we're just making up something that looks right rather than using real data. By contrast, if we damage the surface of a hologram, we just have to change our viewpoint: All of the information is still there to be discovered. We can see this illustrated in Fig. 5-2.

To understand the unique nature of holograms, let's take a step back and look at the physics of waves to inform our explanation of holography.

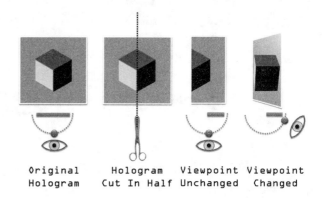

Figure 5-2 *Every part of a hologram contains information about the whole image.*

The Physics of Waves

If we drop a small weight in a fish tank, it will disturb the water's surface, and from the point where the pebble touched the water, we will see a series of concentric waves. These waves will spread out radially in a series of concentric circles (Fig. 5-3).

In physics, we come across waves in all different forms: sound waves, changes in air pressure that take place very quickly; radio waves, part of the electromagnetic spectrum that we can use to transmit information; and importantly for us holographers, the visible portion of the

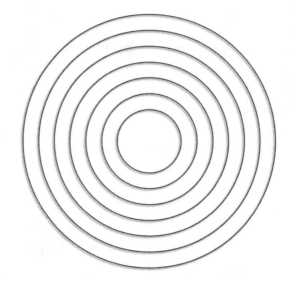

Figure 5-3 *Radial waves.*

electromagnetic spectrum: light waves. We've seen electromagnetic waves in the previous chapter on light and lasers, but let's revisit the concept.

Returning to our fish tank analogy, these waves will move outward from the point at which we dropped the weight at a constant speed, just the same as light waves travel at a constant speed, the speed of light. If we are talking in wave terms, we can call this the *velocity of propagation*. There is also a *wavelength,* which is to say, the distance between ripples. If we imagine a cross section through the fish tank, we'd see the waves consists of peaks and troughs—peaks as the water moves up and troughs as the water moves down. So a better definition of wavelength is the distance between two successive peaks.

Now if I were to drop two weights in the fish tank, with some distance between them, I would see two sets of ripples, as we can see in Fig. 5-4.

These waves would continue to spread out, but where the waves touch, we'd see interference between the two waves. This interference can be constructive (i.e., adding together to combine and amplify the wave) or destructive, where the two cancel each other out. If you can imagine viewing this phenomenon on the screen of an oscilloscope, a device that allows us to visualize waves, you would see the following: To the left we can see our two waves, and to the right we can see what happens when we combine them (Fig. 5-5).

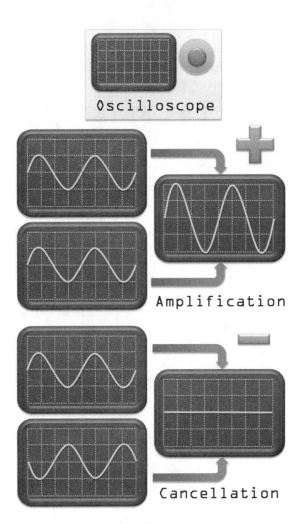

Figure 5-5 *Constructive and destructive interference on the screen of an oscilloscope.*

If our light waves combine, and at that time they are "in phase," that is to say, the two peaks reach each other at the same time, we end up with *constructive interference* where the two waves add up to produce a wave that is much bigger. Whereas if the peak reaches the point at the same point as the trough, we get *destructive interference,* which is to say, the two waves cancel each other out. Think of this a bit like the peaks "filling in" the "troughs." They smooth each other out.

If you can understand this concept of wave interference, you're ready to understand the process of holography, so read on. For those wave junkies who just can't get enough of the physics, we're going to revisit waves a little later in the advanced class that follows our explanation of holography.

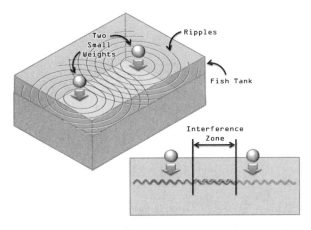

Figure 5-4 *Dropping two weights into a fish tank.*

The Science of Holography

A hologram first of all requires two beams of coherent light; one of these we call a "reference" beam and the other an "object" beam. As we've seen already, these beams interact, and at the point at which they cross, an "interference pattern" is formed (Fig. 5-6).

If we look at that coherence pattern close up, we can see bands of light and dark that form where the beams intersect; of course we can't see this in real life, but imagine being able to see on the microscopic level, as in Fig. 5-7.

Two beams of coherent light intersect in space

Point Source

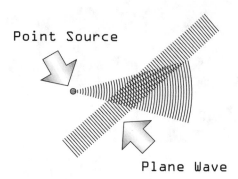

Plane Wave

Figure 5-6 *Two beams of light cross to form an interference pattern.*

It is exaggerated how the two intersecting beams create an interference pattern of light and dark

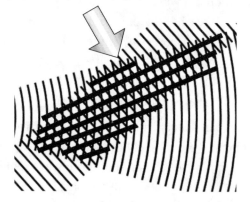

Figure 5-7 *Close-up of light beams interfering.*

In Holography, we capture this interference on a film or a plate that is sensitive to light.

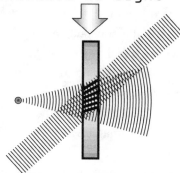

Figure 5-8 *Capturing the interference pattern using a light-sensitive plate.*

By placing a light-sensitive emulsion at the point at which these two beams interfere (as in Fig. 5-8), we can capture the interference pattern on a holographic plate or film. If you look at the chapters on holography chemistry, you can get a handle on how the light-sensitive crystals in the plate work.

Once our hologram is processed, we're left with a record of the "interference pattern"; we've got no information about the reference and object beams themselves, just how they interacted. Take the lasers away, and we're left with something like Fig. 5-9, a piece of holographic emulsion that has recorded an interference pattern.

Only the information about the interference of the beams is captured on the plate or film.

Figure 5-9 *The interference pattern is recorded on the holographic emulsion.*

When a beam is shone at the same angle as the original reference beam, the "object beam" is recreated.

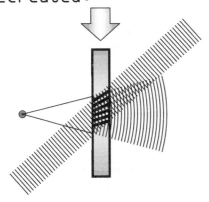

Figure 5-10 *Shining a reference beam through the hologram recreates the object beam.*

This is OK because although the object beam creates a complex pattern, depending on how the light has been reflected from the object, the reference beam is fairly simple. It's just a spread laser beam.

This means that with our laser light; we can "replay" the hologram by shining the reference beam through the hologram, to recreate the object beam. From the hologram will spring a 3D image that recreates the paths of the object beam light rays (see Fig. 5-10).

So we can think about our interference patterns as helping us understand how holograms can capture the dullness of a plastic object—say a chess piece, compared to the sparkle of something shiny. Take a look at Fig. 5-11.

Specular radiation is the name we give to the sort of reflection where we can see an image crisply

reflected in the object. Recall back to the optics chapter our discussion on the laws of reflection. With a specular reflection, the light bounces off the object at an angle where the angular distance between the angle of incidence and the normal and the angle of reflection and the normal are the same. Whereas when we look at a chess piece, it reflects light in a diffuse manner, giving diffuse radiation. As such when light reflects off a chess piece, it is scattered in different directions, and the scattering of this light is recorded on a holographic emulsion. See how a hologram of a piece of diamond or cut glass sparkles like the real thing, and think of the "specular" reflection of the object. In real life, if we look at a diamond from one angle relative to the light shining on it, we will see a bright sparkle, whereas if we change our position, the diamond ceases to sparkle. This is recorded on a hologram because remember we are recording the phase and amplitude information. From view A, the object beam for that point in 3D space summed over the area of the hologram and looking from that particular angle is strong, and it combines with the reference beam waveform to create high contrast in the interference fringes. We get a bright sparkle of an image. Whereas when we change our viewpoint to one where there is no sparkle, the reference beam will still be the same, but the object beam for that point will be weaker, and as such they sum to produce a "dull" image, as we see in Fig. 5-12.

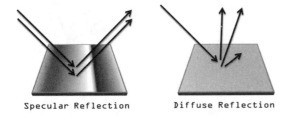

Figure 5-11 *Specular and diffuse radiation.*

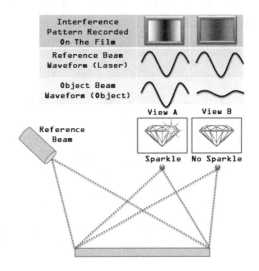

Figure 5-12 *Why a hologram of a diamond sparkles.*

The Physics of Waves (Advanced Class)

Let's return to our fish tank analogy. We've seen a basic explanation of how the waves interact, but where we've simplified this explanation is in our understanding of the object beam. We're got our reference beam as a plane wave, but so far we've done little to explore their interaction in any great detail.

Imagine dropping the two small weights in the fish tank. Remember that we produced two concentric circles. We talked about these waves interacting to produce an interference pattern, well we looked briefly at this interference, but we didn't take time to analyze the *shape* of this interference. Our concentric circles in Fig. 5-13 are the same shape and size, so the "frequency" of the waves coming from the two dropped weights is the same.

Let's draw some lines to join up the points of interference. You will notice that the lines form a series of curves, and the line in the middle is straight, whereas the lines on either side progressively bend more. These curves are *hyperbolic* (Fig. 5-14).

We've seen that hyperbolic curve is a special word for "bent line," but what makes it so special to warrant its own name? If you take a look at Fig. 5-15, all will be revealed. The special property of a hyperbolic curve is that every point

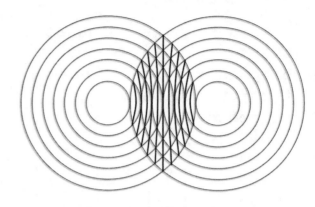

Figure 5-14   *Two radial waves interfering with lines showing hyperbolic curves.*

on the line if connected to two points has a distance from each point that is equal to a constant, so in the case of the lines to the left of the curve, this is 70, and for the lines to the right of the curve, this is 30—plus a difference that is added to the lines on both sides.

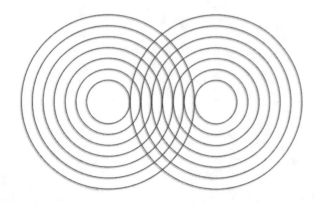

Figure 5-13   *Two radial waves interacting.*

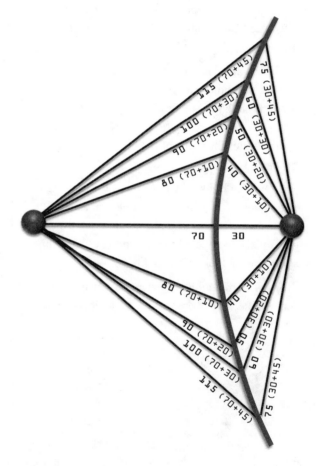

Figure 5-15   *Describing a hyperbolic curve.*

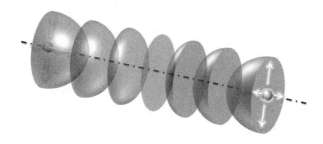

Figure 5-16 *Visualizing in three dimensions.*

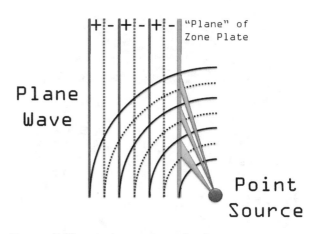

Figure 5-17 *A plane wave and point source interacting.*

We're trying to convey information about 3D imaging, though, in boring old 2D diagrams. So try and think of what is happening a bit more like Fig. 5-16. Rather than a point in 2D space, what we actually have are two points in 3D space, with the interference pattern looking like a series of spherical sections with a flat plane in the middle. If you like, think of it as a series of "bowls" that get progressively shallower with a plate in the middle! If we were to take a cross section of these bowls in a plane that joins the two points, we'd get something that looked similar to the hyperbolic lines shown in Fig. 5-14.

Now imagine light from a laser being spread by a lens. It forms a cone shape, but imagine this cone being formed out of round bowl-shaped slices. If we had a piece of string, we could hold it at the lens and trace a spherical section with a given length of string, and by extending the string, bit by bit, we could progressively describe larger spherical sections.

Imagine that the object we are trying to make a hologram of is a "point" facing the laser, but with its light coming in the opposite direction. Take that imaginary bit of string, and first map out an imaginary sphere holding the string with a length of 1 cm and then gradually extending it to map out successive spheres. Now wouldn't you say that the points at which these imaginary spheres overlapped and "interfered" looked rather similar to our bowl analogy in Fig. 5-16?

We take our "plane wave" reference beam and interact it with a "point source"; as we can see in

Fig. 5-17, something intriguing happens: We end up with an interesting pattern. You'll notice, if you look at the angles to the right of the picture, that we're ending up with a pattern not of identical angles but of angles that change.

We can see in the zone plate that as we get further away from the center of the zone plate, the fringes are packed closer together. Interestingly, if we draw a radial line from the center of the zone plate outward and measure the period of each band, we find that the lengths of the periods are proportional to the square roots of their diameters. Thus when we make a hologram, what we are effectively doing is overlaying a large number of these "zone plates" on the nano scale. Each zone plate represents a tiny point of light on the image, and the interference between them generates the pattern on the film (Fig. 5-18).

If you want a more advanced explanation of how both "phase" and "intensity" information are encoded into a single intensity profile, consider the following. Imagine that Fig. 5-19 is an intensity cross section through our zone plate. It is a function known as sinc (say "sign-cee") and represents the intensity distribution through the center of a zone plate.

When we overlay our reference beam, it has the effect of squaring the sinc function. You'll notice that if you look at the sinc^2 function (in Fig. 5-20), you will see that all of the graph is above the x-axis and there are twice as many ripples. This is

Figure 5-18   *Fresnel zone plate.*

because the intensity information has been combined with the phase information. For those of you who like a bit of math, by squaring this "sinc" function, we get the following graph. What you will notice is that this graph has twice the frequency of the preceding graph. This represents the "intensity distribution" that we see on the zone plate. It combines the information about the "object"—in this case our sinc function, with a coherent background reference. We can see that twice as much information as by removing the reference information from the sinc squared function, we can produce both intensity and phase information.

Transmission versus Reflection Holograms

In this book, you are going to hear a lot about transmission and reflection holograms. Think about them if you like as two different families of holograms. Within each family, there are lots of family members, each a little different with their own little quirks; however, the members of the family all share common traits! What makes them different? Take a peek at Fig. 5-21.

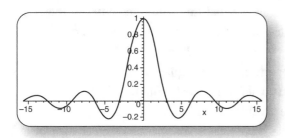

Figure 5-19   *Graph of sinc.*

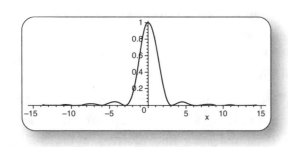

Figure 5-20   *Graph of sinc².*

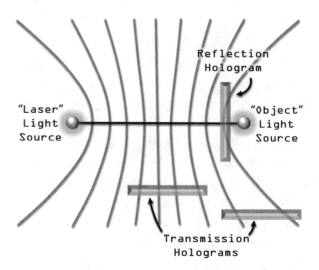

Figure 5-21   *What differentiates reflection and transmission holograms.*

Remember our "bowl" analogy. Let's think of a cross section through these bowls and transpose this onto Fig. 5-21, and we see our familiar hyperbolic lines. Now imagine our reference beam is coming from the left and our object beam is coming from the right. Where we place our plate in relation to the interference fringes generated by our two light sources will determine how these fringes "cut through" our holographic plate. Remember that the light-sensitive emulsion in our plate is going to record these fringes as a series of light and dark patterns.

In the case of a transmission hologram, these slices are going to cut through the thickness of the plate. By contrast, with our reflection hologram, these fringes are going to cut through the width and height of the plate. You can see this in Fig. 5-22.

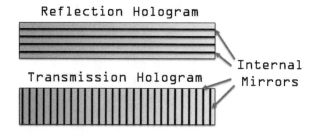

Figure 5-22 *How interference fringes "cut through" a holographic plate.*

These interference fringes manifest themselves as traces of silver once the hologram is developed. In between these silver traces, we find clear lines where there is no silver. Light can pass through this part of the hologram while light hitting the silver is reflected—with the silver lines of the interference patterns acting like lots of microscopic little mirrors!

Chapter 6

Holographic Chemistry

In this chapter we look at the basics of holographic chemistry. Our approach is practical and focuses on the process you need to produce successful images from plates and film. You should be able to read just this chapter and make successful holograms. However, once we've worked through the different ways we can set up our holography lab, we'll go on to more advanced holographic chemistry and indulge in a little explanation of what is actually happening when you immerse your plate or slides in the dishes of chemicals.

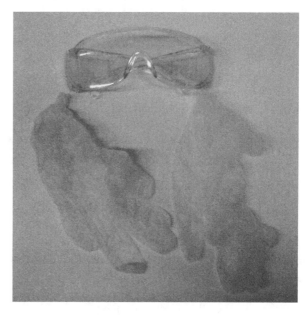

Figure 6-1 *Personal protective equipment is a must.*

Warning

First of all, it goes without saying, that before you even attempt to embark on any chemistry-type activity, you should ensure you have the correct personal protective equipment to do it safely (Fig. 6-1). At a very bare minimum, this should include a set of rigid lab goggles to prevent any splashes from getting into your eyes (you only get one pair of eyes, and you need both of them to view three-dimensional images correctly, so make sure you look after them both!) and a pair of protective gloves to make sure you do not splash chemicals on your hands. Dress for the occasion too. If you know you are going to be working with chemicals, make sure you are wearing a long-sleeve top and long pants that do not leave any part of your body vulnerable to chemical splashes.

You are going to be mixing up chemicals that are supplied in a powder form; some of these are particularly nasty. It is highly advisable to procure a dust mask to use when preparing some of these formulas because inhalation of the chemicals could prove damaging to your health.

You will want to ensure you have a pair of scissors on hand for opening the sachets of chemicals if you are using the Photographers' Formulary kits. I don't know about you, but it's a particularly bad habit of mine that if I can't open a plastic packet and I don't have scissors on hand, I occasionally resort to using my teeth to get through the packet. This is one occasion when you *do not want to use that approach.* Do not bring any of these packets of chemicals anywhere near your face, eyes, nose, or mouth.

As more and more people switch to digital processing, it is harder and harder to source darkroom equipment. Unless you have a local photographic dealer, you may need to look further afield or use kitchen implements as substitutes. This is fine; they will do the same job just as well. However, take a permanent marker, and with any plasticware, measuring jugs, mixing jugs, and mixing implements that you use, make sure to mark clearly that these are "not for food use." When it comes to developing trays, you want a tub that will allow you to immerse the holographic plate in the chemical solutions; however, you also want to ensure that the least possible surface area is exposed to the air, so do not pick a tray that is too big.

You're going to need to mix chemicals with exact quantities of distilled water. If you are making your own chemical preparations, you need to ensure everything is weighed out as accurately as possible. If you are using prepacked chemicals, make sure as you empty out the packets that all of the chemical ends up in the mix and not all over the workspace! Invest in some measuring containers that are accurately marked. If you can't get laboratory or darkroom measures like the ones

Hint

With all of these chemical formulas, it is possible to source the individual chemicals from a friendly college or university lab technician or from a chemical supplier. However, to spare yourself the trouble, the kits from Photographers' Formulary represent good value and are very convenient because all the chemicals are premeasured.

Figure 6-2 *Accurate measuring is vital!*

seen in Fig. 6-2, then good-quality kitchen measures should do the job.

The temperature at which you keep your chemicals will affect the rate of reaction during processing. It is important to keep your chemicals at the temperature recommended by the supplier to ensure that the reaction proceeds at the right rate. You can use a thermometer to check on the temperature of your chemicals (Fig. 6-3). If you need to control their temperature, mixing hot and cold water is not the answer! A far simpler way is to use something akin to the bain-marie (double boiler) you find in the kitchen. Immerse your trays of chemicals in a much bigger tray, which is a "water bath." By adjusting the temperature of this water bath you can indirectly heat and cool your chemicals.

When mixing any of the following preparations, it is essential that you use distilled or deionized water. Distilled water has had all of the impurities removed from it, and is as pure a water as you are going to get. This is important because trace elements in tap water and/or chemicals added to purify the water and improve public health can

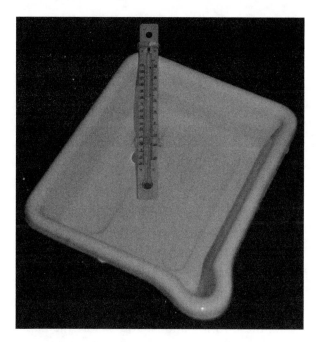

Figure 6-3 *A thermometer keeps track of the temperature of the chemicals.*

Figure 6-4 *Deionized water for mixing chemical preparations.*

affect the chemicals used to develop holograms. If you're wondering where to find this, and if you haven't got a friendly lab technician with access to a supply of distilled water, then you might want to try your local garage or auto parts store. They commonly sell distilled water for maintaining vehicle batteries and the like. If you buy a gallon container, as seen to the left of Fig. 6-4, you will be able to mix a full set of liter measures for your holography chemicals and have a little spare for washing your best holograms.

Disclaimer: The instructions that follow pertain to two well-known holography developers that are currently on the marketplace and reflects the current position regarding the hazardous chemicals these developers contain and their use. These are intended as an instructional guide and are no substitute for the manufacturer's instructions or precautions. Ensure that you have read thoroughly all documentation supplied with any holographic chemicals you purchase before their use.

Project 3: Develop Holographic Plates

You Will Need

- JD-4 Developing Kit from Photographers' Formulary (see Fig. 6-5)

Tools

- Measuring jug
- Mixing jug

Figure 6-5 *The Photographers' Formulary JD-4 developing kit.*

- Glass rod/stirrer
- Access to running water for safety
- Protective gloves
- Goggles
- Trays for each of the solutions

The JD-4 Processing Kit is capable of processing Slavich PFG-03M or ColourHolographics BB-640 emulsions.

Developer Part A (To make 1 L)

Metol	4 g
Ascorbic acid	25 g
Distilled water	1000 mL

Developer Part B (To make 1 L)

Sodium carbonate, anhydrous	70 g
Sodium hydroxide	15 g
Distilled water	1000 mL

Bleach (To make 1 L)

Copper sulfate, pentahydrate	35 g
Potassium bromide	100 g
Sodium bisulfate, monohydrate	5 g
Distilled water	1000 mL

There are two chemicals in these instructions that you must pay particular attention to when working with them: sodium hydroxide and metol. Sodium hydroxide is a particularly corrosive strong alkali, and as such needs to be handled with a great deal of respect. Ensure that when mixing these formulations, you handle the bag with sodium hydroxide carefully. The dust can cause severe damage if it is inhaled or ingested, so wear a dust mask and do not get your face close to the packet. Take care when you are pouring the chemical into the distilled water; take extra precautions to ensure that the contents of the packet end up in the solution. If you spill any of the sodium hydroxide, you will want to ensure that you have plenty of water and damp cloths to wipe it up. Do not use a dry cloth. Sodium hydroxide is highly soluble in water; it will readily dissolve into water. Using a damp cloth and plenty of water afterward ensures that you capture all the sodium hydroxide and disperse any residue that remains. Act quickly because sodium hydroxide has the potential to damage surfaces that you are working on. Metol is widely used in photographic developers as a developing agent; however, some people find that when using Metol for periods of time, they may develop an allergy or sensitivity - which often manifests itself as a rash or other allergic symptom. It can affect the skin, eyes, and respiratory tract so if you experience any symptoms consult your physician. It is also highly dangerous to aquatic organisms, so ensure you dispose of it responsibly.

If at any time while using these solutions you get a "soapy" feeling on your hands, wash them immediately. Sodium hydroxide will cause chemical burns on your skin and is corrosive. Even if you feel no pain or irritation, still wash your hands thoroughly. The action of sodium hydroxide is such that as it burns, you will feel no pain, which makes it even harder to detect if you have spilled any on yourself. Make sure there is a supply of clean fresh water that you can use to wash your hands regularly to ensure that if you have spilt any it is quickly dispersed.

To prepare the chemicals, you will need to mix each solution using the quantities of chemicals stated, adding them one by one to the distilled water. Some of the chemicals come in the form of solids or crystals, which will take some time to dissolve. Be patient and keep stirring, until all of the solid disappears. Don't be overenthusiastic with the stirring! Although it might test your patience, it's far better stirring at a reasonable pace than going too fast, spilling chemicals over everything, and ending up with less solution (and possibly an inaccurate formula if you spill half way through).

Hint

Remember that you need to combine parts of the developer A+B just before use; however, once you do this, the chemicals will have a limited lifetime, so work quickly!

Processing Sequence

Figure 6-6 outlines the JD-4 processing sequence, whilst Fig. 6-7 shows the chemicals as supplied in

Figure 6-6   *The Photographers' Formulary JD-4 developing kit process.*

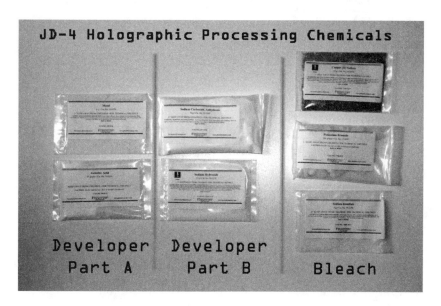

Figure 6-7   *The Photographers' Formulary JD-4 developing kit contents.*

the Photographers' Formulary kit. It will help if once you have mixed your chemicals, and combined the parts A+B of the developer to activate it, you lay your trays out in a logical sequence that allows you to work through this process.

You need to submerge your plate in the developer; try and ensure that it is emulsion side up, to make sure the developer can reach all parts of the emulsion. After about 20 min of agitation, the plate should turn black all over. You should be able to see the safelight through it, but not the bottom of the tray.

Then rinse your hologram for between 20 and 30 s. Use distilled water for best results. Don't be in a rush to complete this stage. If it takes longer, it will be time well spent. Washing for several minutes will make sure all traces of the developer are gone and will leave you with very durable holograms.

Next, move your hologram onto the bleach tray. Keep agitating the tray, slushing it about for a minute or so until the hologram turns clear. Continue past this point for between 10 and 20 s.

Now you will need to rise the plate again; if you use distilled water, you will find you get better results. Twenty to thirty seconds is the lower end of what you should be aiming for. For more durable holograms, give it up to 3 min.

The final step is optional. Prepare some wetting solution, which is a brief squirt of a chemical that reduces surface tension into a tray of water. You only need a drop or two, which will be specified on the bottle. The wetting solution helps you get streak-free holograms as they dry. You can have the lights on now!

Stand your hologram up on its edge with something absorbent underneath. In a pinch, you can use a hairdryer; however, patience often yields better results. If it's a transmission hologram, you should be able to view it right away. For a reflection hologram, things may take a little more time.

Hint

Keep your holograms well away from your sandbox when they are drying as sand will readily stick to the emulsion and make a mess of your holograms.

Some holographers refer to JD-4 as the "JARB" process because it was largely dreamt up by Tung H. **J**eong, Riley **A**umiller, Raymond **R**o, and Jeff **B**lyth.

Project 4: Develop Holographic Film

You Will Need

- JD-3 developer kit from Photographers' Formulary

Tools

- Measuring jug
- Mixing jug

- Glass rod/stirrer
- Access to running water for safety!
- Protective gloves
- Goggles
- Trays for each of the solutions
- Bright light source

The Photographers' Formulary kit (Fig. 6-8) comes as a series of chemicals in plastic bags, which must be mixed in sequence with the correct amount of distilled water in order to make the working solutions.

> Treat all chemicals with caution and respect; there are two chemicals that you will be using here that are particularly harmful:

Potassium dichromate must be stored carefully, as it is an oxidiser; which means that it can release oxygen in a reaction—which can promote fire. It is also toxic. To dispose of the solid, you should dilute with large quantities of water;

ensure that you never simply put any of the solid chemical in the bin. This chemical can cause ulceration to the skin as a result of chemical burns. Compounds of chromium are known carcinogens, so you should avoid exposure to them. You should ensure that you treat this chemical with the utmost respect; wear rubber gloves and personal protective equipment at all times and ensure that you clean equipment that comes into contact with it with copious amounts of water and soap and water.

Sodium bisulphate used in this formulation is particularly nasty. It should not be swallowed. If it is ingested, seek urgent medical attention and do not induce vomiting. Any spills on skin should be washed under a stream of running water for a quarter or an hour or more. When you mix the solution, ensure that you wear full personal protective equipment including rubber gloves, apron, dust mask, and safety goggles.

Catchetol will severely irritate sensitive parts such as eyes, skin, and mucous membranes. For this reason ensure full personal protective equipment is used when mixing solutions. It also has the potential to poison the liver and the kidneys. It is also a central nervous system depressant and methemoglobin former, which is to say, it will form compounds in your blood that will not absorb oxygen. Therefore treat this chemical with a great amount of respect.

The Urea and Ascorbic acid in this formulation have the potential to irritate skin, eyes, and mucous membranes.

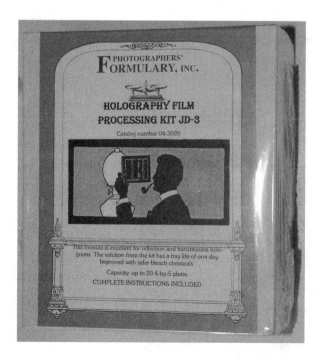

Figure 6-8 *The Photographers' Formulary JD-4 developing kit.*

Developer Part A (To make 1 L)

Catechol	20 g
Ascorbic acid	10 g
Sodium sulfite	10 g
Urea	75 g
Distilled water	1000 mL

Developer Part B (To make 1 L)

Sodium carbonate, anhydrous	60 g
Distilled water	1000 mL

Bleach (To make 1 L)

Copper sulfate	17 g
Potassium bromide	55 g
Succinic acid	2 g
Distilled water	1000 mL

Posttreatment (To make 500 mL)

Ascorbic acid	10 g
Distilled water	500 mL

To prepare each of these solutions, you will need to dissolve each of the chemicals in turn, in the requisite quantity of distilled water. Some of the chemicals are tricky to dissolve and may take some time. You will want to ensure that you have a suitable stirring implement—something that is inert and will not react with the chemicals, and definitely not your hands! As you will want to see how much of the solid chemical has dissolved, I strongly suggest that you mix them in a glass jug so you can see how much more stirring there is to go! Also, as you will need to stir quite vigorously. I strongly suggest that you pick a container to mix the chemicals where they are unlikely to splash over the edge.

Hint

Remember that you need to combine parts of the developer A+B just before use; however, once you do this, the chemicals will have a limited lifetime, so work quickly!

Processing Sequence

Figure 6-9 outlines the JD-3 processing sequence, whilst Fig. 6-10 shows the chemicals as supplied in

Developer A+B → Rinse → Bleach → Rinse → Post-treatment → Wetting Solution

Figure 6-9 *The Photographers' Formulary JD-4 developing kit process.*

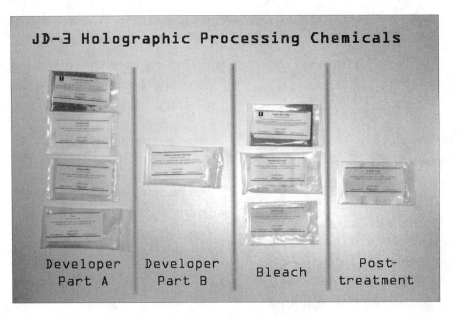

Figure 6-10 *The Photographers' Formulary JD-4 developing kit contents.*

the Photographers' Formulary kit. It will help if once you have mixed your chemicals, and combined the parts A+B of the developer to activate it, you lay your trays out in a logical sequence that allows you to work through this process.

JD-3 takes a little while longer than JD-4, so you can expect to wait around 2 min for all parts of your hologram to turn fully black.

As with JD-4, you can rinse for 20 to 30 s at the minimum; however, allowing for a few minutes will mean that you produce more durable holograms in the long run.

You can place the hologram, once it is rinsed, into the bleaching solution. This will require around 2 min for the plate to turn completely clear—again longer than JD-4.

The hologram will then need to be rinsed a further time; again, short times will result in inferior results. Half a minute cuts it fine; a few will produce a more durable finished product.

There is one additional step with the JD-3 process, the posttreatment bath. You will need a bright light source above the tray. Soaking the hologram in this bath will turn it from a pink to a brown color.

The final stage is the wetting solution, which is the same as with JD-4.

Reflection holograms require thorough drying, but transmission holograms can be viewed right away.

Hint

If you are having a problem identifying the emulsion side of the film, there are two methods. A slightly moistened finger should be able to feel on the very edge of the film that one side is slightly "tackier" than the other. Alternatively, on 4 × 5 sheets of film, there is usually a notched or cut corner. If you hold this notch in the top right-hand side, then the emulsion side of the film will be facing you (Fig. 6-11).

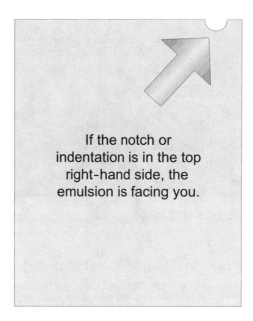

If the notch or indentation is in the top right-hand side, the emulsion is facing you.

Figure 6-11   *The notch in the top right-hand side.*

General Instructions for Developing Holographic Plates/Film

These instructions apply regardless of what formula you are using to develop your holographic materials.

When it comes to developing your holograms, you are going to need some small trays that you can use to immerse your holograms in their developing chemicals. Plasticware, photographic trays, or any small receptacle that is slightly bigger than the holographic materials you are working with will do. You need to ensure that the material is inert and will not react with the chemicals, so no metal, painted surfaces, or anything with a design that could wear off. Just plain old plastic or glass works best. It's a wise practice to have a sink nearby. Even though you will have mixed your chemicals with distilled water, it's still good to have a supply of water in case of spills or getting any onto your skin. An example of how you might set up your trays is shown in Fig. 6-12.

Figure 6-12   *A small hologram processing setup.*

The house that I moved into while writing this book is an old country house that contains its own photographic darkroom in a cupboard under the attic stairs. Already plumbed in was a small sink; and there is sufficient space to place a couple of trays. However; if you haven't got space for a dedicated darkroom, the next best thing is to try and find a small room in your house that is dark and can temporarily be taken over for a weekend or week to make some holograms. If you have a second bathroom, this is probably a good space to start. With a sheet of plywood and some blocks of softwood, you should be able to fashion a board that will temporarily fit over a bath or sink to provide a level work surface for hologram processing.

Figure 6-13   *Plastic tongs.*

Tongs are your friends! They make the process of moving your plates from tray to tray really simple. Look for cheap tongs made of plastic (Fig. 6-13). These will do the job; however, they can sometimes scratch the emulsion on the surface of your holographic plates. Better photographic tongs have rubber ends, which serves to protect your plates a little against scratches, and it also helps grip the plate more effectively!

Storing Chemicals

Once you have mixed up the different solutions, they will store well individually; however you need to take some precautions. Find some 500-mL bottles. I find that I divide my liter of chemicals into four portions. I save half a liter of each in a bottle, and then I mix 500 mL of developer from parts A&B, keeping the other 250 mL of each in reserve for when the developer is exhausted. This way you can have "four sessions" of holography from one mix of chemicals. You want to ensure that there is as little air in the bottles in which you store your chemicals as possible. There are a couple of ways of doing this. You can take small inert objects (e.g., glass marbles) and drop them into the neck of the bottle to make the level of the chemicals rise to the point where you have excluded most of the air from the bottle. Another option, from a photographic supplier, is special bottles that are designed to "concertina." The concertina allows the bottle to expand and contract, changing its volume to suit the fluids that you want to seal inside. By concertina-ing the bottle in and out, you can make a bottle of an appropriate size to store all of your liquids and exclude all air from the bottles— preventing your chemicals from oxidizing. This way you can vary the volume of the bottle to suit the contents; squeezing the bottle helps you to exclude air. Of course, you can squeeze ordinary plastic bottles, but they will not retain their shape and may be hard to store or prop up on a shelf.

Sometimes if you are pressed for space; or for example if you need to expose your holographic plates and develop them in separate places, a darkroom bag is really useful. It's a bit like the kids' game where objects are placed under a blanket and you have to guess what they are by feeling. A darkroom bag has two elasticized armbands designed to try and restrict the amount of light that enters. There is also a zipped pocket for your holographic plates or film. Once everything is in the darkroom bag, you can switch the lights on and work with it in the bag. Figure 6-14 shows a darkroom bag.

Figure 6-14   *Darkroom changing bag.*

Waste Disposal

Never tip chemicals down the drain. If you are going to be a holographer, make sure you do so responsibly! Science should enhance the world that we live in, not make it worse!

Contact local authorities and ask them about the correct procedures for handling hazardous waste. The chances are that your local civic amenity center or refuse handling center has facilities for handling such low-level chemical waste.

The easiest way to manage is to buy your distilled water in gallon containers. Once you have finished with your chemicals, tip them using a funnel into the containers for later disposal. Remove the labels and make sure you put on a replacement label that states the contents are chemicals that are to be treated as hazardous. These can then be safely stored and transported for disposal.

Chapter 7

Your Holography Workshop

Your holography workshop is the place where you are actually going to make your holograms. We've discussed some of the basics of how you can mount your optics, but we've still not gotten to a point where you can do anything with them. Unless you are making the simplest of holograms, you will want more flexibility than working with a simple optics bench.

Choosing a Site for Your Holography Workshop

Give careful consideration to the site where you will be making your holograms. If you can find a basement or a ground-floor room with a concrete floor, this provides by far the best option from the point of view of stability of your setup.

Try and avoid building your holography setup on a first floor with a wood floor, especially one with creaky floorboards. These are much more susceptible to vibration, and any movements may cause your table to wobble.

Consider also if there are any nearby roads over which trucks pass or railway lines where trains thunder by. Try and identify sources of vibration or noise that could cause your table to wobble, and site your holography workshop away from these because the smallest vibrations can ruin your holograms!

If you're restricted by where you live and your only option is to build your holography workshop in a room adjacent to a railway, then at least get a copy of the timetable of when the trains run so you know when not to make holograms.

Project 5: Construct a Sandbox

You Will Need

- 20 mm/3/4-in plywood
- 100 mm/4-in-wide planed sanded softwood
- Bicycle/wheelbarrow inner tubes

Tools

- Drill
- Screwdrivers
- Hammer
- Hand saw/jigsaw/circular saw

In this project, rather than be didactic and specify a set of measurements, I've left it open to your interpretation because I'm guessing there will be a vast array of different amateur holographers out there with very different spaces available and different means.

You will want three sheets of plywood: one for the base of the sandbox, one to sandwich a series of inner tubes between, and a third to sandwich some thick carpet between to add an additional layer of protection against vibration. If you cut an 4 × 8-ft sheet of plywood into three sections and have a comfortably sized holography box, with care you will be able to accommodate all the setups in this book if you are using small plates (Fig. 7-1).

You will need to frame the top sheet of plywood to contain the sand. My own preference is for planed sanded softwood because it is pleasant to work with, easy to cut, and won't give you splinters when you are using the sandbox if your wrist happens to catch the edge. You could equally use plywood from the same sheet, although take some time sanding the edges after they have been sawn to ensure the table won't give you splinters.

If you want to be really smart, grab a black permanent marker and create a "black white black white" pattern at 100-mm/4-in intervals around the perimeter of your sandbox. This will help when

Figure 7-2 *Rubber inner tube.*

you draw diagrams of your holography setup for later reference because it will provide a built-in "ruler" for your holography table.

You will need a number of inner tubes to support your sandbox (Fig. 7-2). Because the inner tubes are filled with air, they help to isolate the sandbox from the rest of the installation, preventing vibrations from reaching the sandbox. The sandbox literally rests on top of these inner tubes: There is no other physical connection between the sandbox and the area on which it is mounted. Do not underestimate the weight of the sand in the sandbox! Although the weight is spread over a number of inner tubes, you should ensure they are all partially inflated to the same pressure so the weight is evenly distributed across the table.

Optionally, beneath your inner tubes, you may want to add another layer of thick carpet and or felt, sandwiched between another sheet of plywood, which will also help to deaden the vibration.

Next you will fill your sandbox with some sand. The sand fulfills a number of functions. It provides a medium to support your optics, but it also adds weight to the box and helps to damp

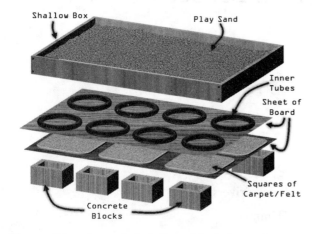

Figure 7-1 *An exploded view of the holography sandbox.*

Figure 7-3   *Play sand.*

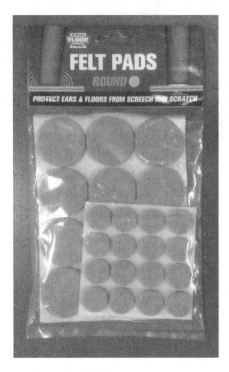

Figure 7-4   *Felt pads protect surfaces.*

vibrations. Because sand is granular, in the event of any vibrations, it helps to isolate them from the optical components in your sandbox. Although you may be able to procure sand from a builder's supply, by far the most pleasant sand to work with is "play sand" designed for children's sandboxes (Fig. 7-3). It has been presifted to ensure there are no bits of stone, and it is fine sand that sticks together and does not leave color on your hands.

The final sheet of plywood will need a solid surface to rest on; if you have a concrete floor, this is ideal because it will provide a strong mass that will damp vibration. If not, source some heavy concrete blocks or paving slabs that will fulfill the same role.

If your sandbox is going to be placed temporarily on a delicate surface, say for instance you are building your sandbox over a bath, you may want to consider adding some felt pads under the final sheet of plywood to protect the surface underneath (as shown in Fig. 7-4).

Here's another option worth investigating. If you go to a high-end audio shop, you can find "cones" turned from metal, which are designed to screw to the bottom of high-end loudspeakers. These sit inside a little metal cup. The weight of the sand will ensure there is significant weight on the points of these cups, so be careful what sort of surface you rest them against. But they do provide another tool in the holographer's armory in the war against vibration.

When you are finished, you should be left with a smooth, enclosed sandy surface such as that shown in Fig. 7-5. Keep a sheet of card handy for smoothing the surface of the sand in between setups.

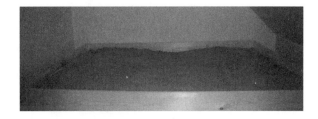

Figure 7-5   *Surface of the finished sandbox.*

Figure 7-6   *An example of an interferometer setup.*

The optical components can then be simply pressed in the surface of the sand as shown in Fig. 7-6. It is no accident that we have picked an interferometer setup to illustrate the holography box in use. At this stage, it might be worth fast-forwarding to the chapter on science fair projects and building the interferometer setup. This will help you identify sources of vibration and also help you ascertain how long after being "vibrated" your holography sand table takes to relax to a steady state.

Portable Sandboxes

Portability and stability do not necessarily go hand in hand. If you make a holography table that can be wheeled around, it won't necessarily be the most vibration-proof table. However, in some circumstances you may not have the luxury of being able to set up a holography table in one place for a number of weeks. Therefore, consider adding casters to the bottom of your table with a temporary clamp that can hold the "layers" of the sandbox table together when you are moving it. The layer of sand in the sandbox will weigh more than you think, so take care in construction to make a sturdy box, or you may end up with sand going everywhere!

Making Your Room Lightproof

You need to ensure you have a dark area in which to conduct your holography setups. Houses often have a second bathroom that you can monopolize for your holography setups, and often these bathrooms are built in the center of the house or under the stairs where there are no windows. If you can find such a room, you have an ideal space for a holography setup. However, life is far from ideal, and often the only room available will have a window.

You need to find a way to "block out" the window so no light can penetrate. An easy way to do this is to build a frame of some lightweight section softwood that fits snugly within the window reveal. Then find some opaque black plastic and staple it over the frame, ensuring that as you do so, you wrap it around the frame. This should then fit snugly into your window reveal and block out the sunlight. If the design of your window is such that it doesn't naturally hold itself, you may want to consider a few judiciously placed hooks inconspicuously screwed into the wall or window frame, with an eyelet on the blackout cover to hold it in place. Velcro is another good temporary solution if you want something that will hold but then be easy to remove.

Once you have masked out the main source of lights, take a look around in your "darkroom" to see where else light might be creeping in. You may find that light seeps in around the frame of your doors. In this case, install some draught excluders; these black plastic brushes and foam strips will not only help keep the light out of your darkroom, but they'll also help the thermal performance of your house!

And here is a key point: Try and isolate your holography setup from any sources of heat fluctuation. Drafts are one example; radiators and heater are another. You want to make holograms when the heat is switched off! A gust of warm or cold air of a different temperature to the ambient has the potential to upset your holography setup by disturbing the fringe patterns. We're dealing with things on such a small scale that all of these factors are important.

You might also find that you have light leaking in through vents or holes in the wall, in which case you will need to fashion some temporary covers to prevent light egress.

Safe Lighting

The emulsion that we will be using in this book to make holograms is "red sensitive," which means that any red light (including white light that has red as a component) can "fog the emulsion." For this reason, we need a safe light (like the one in Fig. 7-7; that is to say, a light that will not fog the emulsion).

In this case, we want to aim for a dim green light. There are a couple of ways you could approach this problem: Using a simple LED you can make a safelight. Just ensure you select a green LED (or tricolor LED that can be switched to a different color if you work with green-sensitive emulsion) or you can use a piece of green gelatin plastic (Fig. 7-8) to mask over an existing light.

Figure 7-8   *Green gelatin filter.*

Sundries for Your Holography Workshop

To make bright holograms, oftentimes you will want to enhance the reflectivity of some objects and darken out backgrounds or supports that are not meant to be seen. To this end, it makes sense to procure some white and black hobby paint, as shown in Figs. 7-9 and 7-10. Keeping these on

Figure 7-7   *Safelight constructed from small torch and filter.*

Figure 7-9   *White spray enamel.*

Figure 7-10   *Black spray enamel.*

Figure 7-11   *A packet of poster adhesive.*

hand will be useful when you are building models to create holograms from.

Supporting Your Subjects

To make good-quality holograms, your subjects need to be rigidly supported so they do not move. To this end, two supplies are worth keeping on hand: poster adhesive and polymorph.

Poster adhesive in the form of a sticky clay that can be rolled into balls is commonly sold under brand names such as Blu-Tack (Fig. 7-11). It is useful because it is mildly adhesive and can be shaped to support and hold holography models in place. Do not be too frugal when using it to stick your models to supports because the more rigidly your models are supported, the better your chance of success.

Polymorph is less well known but a fantastic material that every hobbyist should know about (Fig. 7-12). Effectively it is a hardwearing durable

plastic with a relatively low melting point. You can pour a cup of boiling water, tip in some polymorph granules, fish them out because they will stick together in a large mass, and then working with your hands (although it is a little hot), you can fabricate a variety of different shapes, even pressing your subject into the polymorph in order to take a form. As the plastic mass cools, it forms something akin to a very hardwearing nylon plastic

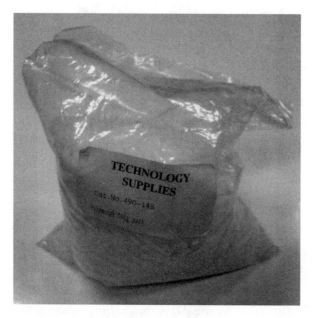

Figure 7-12   *A bag of polymorph.*

that is durable and rigid. In addition to securing models, it's also invaluable when building optical supports.

Timers

To get consistent results from your holography setup, it is useful to have a small timer on hand. You can use it to time hologram exposures and also ensure that you develop your holograms in accordance with the timings supplied with your chemicals. Using consistent timing and making notes about what works and what doesn't will allow you to refine the art of developing your holograms to the point where you find a formula that will produce consistent results.

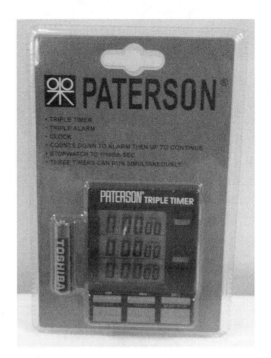

Figure 7-13   *Light meter.*

Light Meters

It's useful to have a light meter to help you gauge the incident light that falls on your hologram and to help you determine the relative brightness of the reference and object beams. The light meter in Fig. 7-13 is a standard photographic light meter; the light meter in Fig. 7-14 is a light meter built for a holography course that was run by the Open University along similar lines to the light meter featured in the electronics project section of this book.

Checking Relaxation Time

Once you're all set to go, before you start making any holograms, it is highly advisable to skip ahead to the chapter on science fair projects and set up the "Michelson Interferometer." It uses standard components that you will have bought for your holography setup, so you won't need to purchase any additional supplies. However, it will help you

determine the "relaxation time" for your table, which is how long after an interruption the table takes to settle to a calm state. Do this every time you change something about your table or move it into a different position. Now you are all set up, let's make some holograms!

Figure 7-14   *Alternative light meter.*

Chapter 8

Simple Holography

We're going to start with the most basic holography setups possible. In this chapter we concentrate on using setups that work well with simple optics and holographic plates for ease. I'd recommend that everyone start with plates, even though they are a bit more expensive because they don't have the problems inherent with film of "flexing" and require less hassle to mount.

If you have read through some of the chapters on constructing your own equipment and have decided that time is more valuable than money, I will make a couple of recommendations. Integraf, listed in the suppliers' index, sell two holography outfits, their "Standard" and "Student" holokit, that are supplied with a diode laser, holographic plates, and chemicals to develop plates. The difference between the two kits is the price and the number of plates you get. "Industrial Fiber Optics" is also listed at the end of this book. Meanwhile, buy a kit that comes with film, the chemicals to develop film, and a really nice set of starter optics ideal for use in a sandbox, as well as some off-the-shelf equipment that can be used for some of the advanced experiments in this book (for which a helium-neon laser would be a good investment). Because one kit comes with film and the other with plates, one comes with a diode laser, and the other recommends a helium-neon laser, they really are quite complementary. And if you were to purchase both, with a few additions, you would have the requisite supplies to tackle most of the projects in this book.

Project 6: Direct-Beam Reflection Hologram

Hint

In this project, we are going to go through all the basics of making a hologram, taking some of the theoretical lessons learned in previous chapters and applying them in practice. So be sure you work through this project before tackling any others in the book. It will give you a good grounding in practical holography and set the scene for more complex experiments.

You Will Need

- Laser
- 2x D cell batteries (nonrechargeable—for 1.5 v!) and holder (only if you are using a laser diode)
- Lens (which you won't need if you use a laser diode)
- Plate (Slavich PFG-03M emulsion)
- Object
- Holography processing chemicals, etc.

For all of the projects in this book, where "lens" is specified, you want to use a divergent lens, with a focal length of –10 mm or less.

A holographic plate consists of a sheet of thin glass with a coating of light-sensitive emulsion. If you are on a tight budget, you can buy a small box of six plates that are 2.5 × 2.5-in each as shown in Fig. 8-1. If all you want to get out of this book is a quick play around and the feeling that you have "made a hologram," your cheapest way of doing it will be to buy six plates, a set of chemicals, a laser diode, and find a couple of tubs for processing. If you can find a vibration-free surface and in a pinch get away without the use of a sandbox, then you've got the chance to make holograms and have a great night in, for not too much more than a decent night out!

However; I am sure that when you look at all the exciting possibilities, you will want to delve into this subject in more depth. To begin with, anticipate making many mistakes and ending up with lots of "dud" holograms. If you've got the cash, invest in a box of 30 small 2.5 × 2.5-in plates (shown in Fig. 8-2). When you've got the bug, you'll find it really frustrating waiting for the next

Figure 8-2   *Box of 30× Slavich PFG-03M plates.*

small box of plates to come through the mail, so save yourself the heartache in advance and buy a big box of plates.

When you get even more advanced, you might consider migrating up to plates that are even bigger still. The next size up is the 4 × 5-in plate (10 × 12.5 cm). These allow you to make a decent size hologram that is suitable for display; however, the downside is that they cost more than the small plates. You can see these in Fig. 8-3.

Figure 8-1   *Six Slavich PFG-03M plates.*

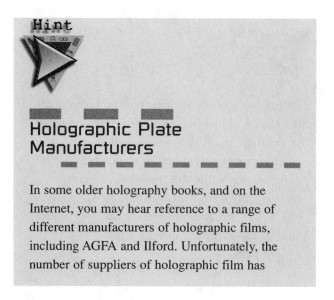

Hint

Holographic Plate Manufacturers

In some older holography books, and on the Internet, you may hear reference to a range of different manufacturers of holographic films, including AGFA and Ilford. Unfortunately, the number of suppliers of holographic film has dwindled, and references to some emulsions such as the classic AGFA 8E56HD holographic emulsions are long out of date. Although there are a handful of manufacturers of specialist holographic emulsions that are used for color holography and specialist applications, the chances are you will be using emulsions manufactured by Slavich; made in Russia, these are relatively cost effective, and there are a fair range of different suppliers. Some holographers have stockpiled old emulsions and kept them in fridges long past their sell-by dates; however, to increase the chances of success, stick with new emulsions brought from recognized suppliers. We use the red-sensitive PFG-03M emulsion for most of the projects in this book that use holographic plates. Where we use "film" rather than plates, we switch to the PFG-01 emulsion. Be mindful that these both require different developers, and the development steps for both are outlined in the chapter on holographic chemistry.

Figure 8-3 *Box of 5 × 4 × 5-in Slavich PFG-03M plates.*

Laser Diodes

If you want to make simple holograms, the cheapest way into the hobby is to purchase a diode laser. These are the type of lasers that you would find inside a handheld laser pen, and they have made the hobby much cheaper because the laser is one of the biggest investments you must make to get into the hobby. Now that investment has shrunk to not much more than a large handful of change.

The laser diode in Fig. 8-4 was obtained from Integraf, listed in the suppliers' index. The benefit of this device over the cheap laser pens you can buy is that this laser diode has an easily removed collimating lens. Another advantage of this laser diode is that it requires no driver circuitry, just a supply of three or so volts that can be supplied by a pair of D-cell batteries in a carrier.

A few other features make this sort of laser diode stand head and shoulders above some of the competition: The frequency output of this laser diode is stabilized, which is essential for good-quality holograms. Just think—if all you're using a laser for is to point at, say, PowerPoint slides in a presentation, it doesn't need to be high quality and frequency stabilized. If the color changes a bit, by an amount that is not discernable by the human eye, so what. However, with our holograms, this can make all the difference. Also, you'll see a discussion of coherence length when we get to

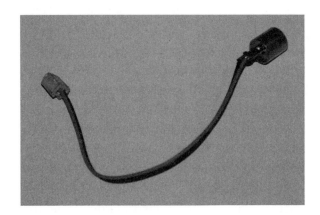

Figure 8-4 *Laser diode from Integraf.*

Fig. 8-17. Note that the coherence length of this particular diode is rather good; in fact, unlike many cheap laser pointers, this laser diode once warmed up for a few minutes can have a coherence length that exceeds 1 m.

One more neat feature: The lead of the laser diode is terminated in a little plastic socket, as can clearly be seen in the picture. While designed to clip into a printed circuit-board plug, this is actually quite handy because it allows you to poke the pre-tinned wires of a battery holder in and they will make a good contact, which can be enhanced with the addition of some tape.

We saw the construction of a diode laser in Chap. 4; from the laser diode element itself, a beam emerges that is oval in shape and spreads out like a cone; however this doesn't tally with our perception of laser pens as producing a thin red line. The reason for this is simple. Manufacturers tend to add to the laser pointer a small collimating lens, which takes the divergent oval beam and "straightens" it to the laser beam that we all recognize. Alas, in holography, we often want a spread beam for simple holograms. Although we could "spread" the beam, with the aid of a few lenses, every time we pass the beam through some optics, we have the potential to make the beam dirty; which will inevitably degrade the quality of our holograms. By keeping the optics as simple as possible (i.e., using the beam directly from the laser) we can produce high-quality yet simple holograms!

In Fig. 8-5, you will see the laser diode from an angle that clearly shows the black plastic collimating lens has been removed. The silver cylinder in the middle of the diode is the actual diode element itself. The construction of the laser diode is rather fragile, especially how the power leads are soldered to the rear driver-printed circuit board. My own recommendation is to wrap a few turns of adhesive tape around the junction between the circuit board and the leads to act as a strain relief and ensure the longevity of your laser diode.

Figure 8-5 *Close-up of the laser diode with collimating lens removed and strain relief added.*

In the diagrams in this book, we have shown the lasers as "helium-neon" lasers, with a lens to expand the beam of light. However, if you are using a diode laser, you just need to ignore this first lens, and imagine that the spread beam that you are using has come from a laser-lens pair.

A plethora of laser diodes are out there; acceptable results can be achieved with cheap equipment, and for the casual experimenter who just wants to try a handful of projects, an inexpensive laser diode will satisfy your curiosity. However, if you want to make quality, repeatable holograms, aim for something with the laser diode specs outlined in the nearby box.

- Integraf Laser Diode Tech Specs
- Class IIIa diode laser
- Output 3–4 mW when powered at 3.0 vdc
- Wavelength 650 nm
- Coherence length: Exceeds 1 m if powered by batteries and allowed to "warm up" for a few minutes.
- Stabilized frequency output
- Removable collimating lens

Simple Holography

Helium Neon Lasers

For some of these simpler holograms, the expense of a helium-neon laser is unnecessary, and we can work perfectly well with a basic laser diode. However, if you just happen to have one knocking around or if you plan to do some more advanced holography and so have bought one anyway, it really is very simple to use. You want to use some lenses to spread the beam with a helium-neon laser; you may find that one lens is insufficient to spread the beam wide enough to manage an exposure on a small table. If this is the case, you can use two diverging lenses aligned one after the other as shown in Fig. 8-6.

Making Your First Hologram

To begin with, let's make an exceptionally simple hologram. This uses a method called "contact copy," whereby your subject is directly in contact with the holographic plate. You can choose a range of different things for your subject, but you want to pick something that is quite flat and very shiny and reflective. A good subject to start with is a handful of shiny change from your pocket. It's a quick and easy subject that is little hassle to set up; however, you can get creative and explore your house for other things that are relatively flat and shiny. You want a subject that is firm and rigid, nothing soft or fluffy! You also need something that is going to reflect red laser light, metallic objects, light objects, and anything of a reddish hue. Avoid anything that is blue or green because it will not reflect the light well. Metal ornaments are sometimes good subjects as are toy cars in the colors just specified and other small novelties. Remember, as shown in the holography workshop chapter, if something doesn't reflect red laser light readily and you aren't too worried about it getting covered in paint; you can always take it out of your workshop and spray it with white or silver paint. In Fig. 8-7, I've used a heart-shaped ladies' mirror. This is a nice subject, as in addition to the reflections that you will see from the shiny chromed surface, there is a heart-shaped ring of cut-glass crystals. In the finished hologram, when it is tilted toward the

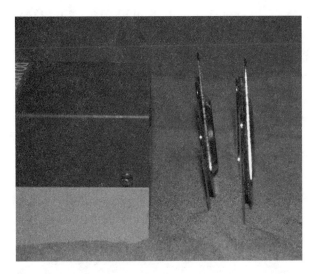

Figure 8-6 *Using two diverging (convex) lenses to spread the beam of a HeNe laser.*

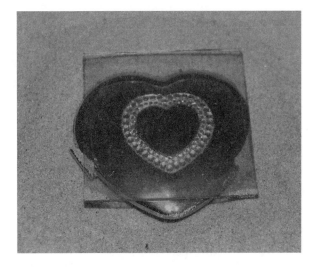

Figure 8-7 *Ladies' heart-shaped mirror used to make contact reflection hologram.*

light, these shine and sparkle like diamonds and create a really impressive effect that highlights why holograms differ so much from other imaging methods.

Also, be wary of plastic objects; plastics, especially thermoplastics, deform readily with changes in temperature. Although this change is imperceptible visually, on the micro scale of our holographic fringes, this expansion can make a big difference, and changes in temperature could ultimately ruin your hologram.

Bad Vibrations

Unfortunately, the Beach Boys had it wrong; when it comes to holography, vibrations are indeed bad! If you skipped ahead to the science fair projects, and got to the point of building the Michelson interferometer, then you will have established the "relaxation time" of your holographic bench and/or sandbox. Keep this time in mind; you will want to ensure that all vibration has thoroughly subsided before making a holographic exposure. Even a tiny vibration as small as a thousandth of an inch has the potential to totally ruin your hologram day. "Bad, bad, bad, bad vibrations" might not have made for such a catchy lyric though.

It's simply a matter of placing your object flat in the sandbox with a plate resting over the top. You will need to position your laser so it shines overhead. If you have a laser diode, this is relatively easy because it is so small and lightweight. With a helium-neon laser and lens, you might want to consider bouncing the spread beam off a mirror because it will be harder to mount the laser itself overhead.

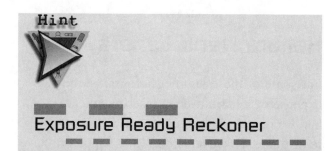

Hint

Exposure Ready Reckoner

If you want to get really heavy about the science of holography exposure, you can give the advanced chemistry chapter a read where we'll tackle the exposure of your holograms in a more mathematical manner. However, for the purposes of this exercise, I'm going to give you some clues that will put you in the right ballpark. If you are using a diode laser, for example, the Integraf diode laser, with a spread beam, if your object is between 14 and 16 in or 35 to 40 cm from the laser, then you want to expose your hologram for around 10 s; however, you can interpret this as at least 5 s, and no more than 40 s, and you should produce something meaningful.

Turn to the chapter on holography chemistry and follow the instructions for processing holographic plates. When you pull your hologram out of its final wash, don't be downhearted if it doesn't sparkle with three-dimensional brilliance at you right away! The emulsion of the plate will swell with the water during the process of, well, processing, and you will need to wait for it to dry before the image becomes clear. Figure 8-8 shows a finished hologram produced using this method.

Figure 8-8 *Reflection hologram of heart-shaped mirror.*

Viewing your First Reflection Hologram

When it comes to viewing your first reflection hologram, there are a few things that you need to remember. You need a point source of light; it doesn't have to be monochromatic light, like a laser. You can use light from a bulb with clear glass or the sun. Have the light source positioned above you and shining from behind you; hold the hologram above you and change its angle with respect to the light. Figure 8-9 illustrates this diagrammatically.

To enhance your hologram's appearance, take some black paint and coat the back of the hologram on the emulsion side. This will help improve the contrast of the image when you view it. If you are feeling more adventurous and have ready access to chemicals, in the advanced chemistry chapter there are instructions on how you can chemically blacken your holographic plates.

Improving the Setup

Although this method might be incredibly simple, it also places some constraints on the sort of subject matter we can use because it has to be flat and capable of lying on the sand. Once you've confidently produced one of these contact holograms, we can start to look at making holograms where the plates are vertically supported in the sand rather than lying on top of it.

In Fig. 8-10 you can see how this is practically accomplished. A combination of a piece of black tubing and a black clothes pin are used to support the laser diode in the sand, and a couple of large stationery clips are used to support the holographic plate.

Figure 8-11 shows us a diagram from the side of this type of setup. Remember, the closer you position your object to the holographic plate during exposure, the brighter the image of it

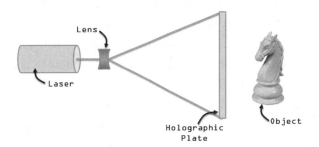

Figure 8-10 *Photo of a more advanced single-beam reflection setup.*

Figure 8-9 *Viewing your first reflection hologram*

Figure 8-11 *Diagram of single-beam reflection setup.*

will be. The beam produced by your laser will follow a "Gaussian distribution"; if you make the beam so that it is around a third larger than the plate you are using, you will ensure you use the brightest portion of the beam to make your exposure.

Using a Shutter Card

Now it's all well and good using our hands to block the laser light with a simple hologram. But as we progress, it's wiser to use a shutter card because we can ensure that the light is blocked out more thoroughly, and furthermore, the card can be left in place while we maneuver our emulsion into position.

I like to use a "dummy plate" that can either be a failed hologram or a piece of white card the size of the plate you are using to set up the hologram. This allows you to get the laser in place and all the optics adjusted to ensure you are making an exposure over the whole of the plate. Then you can use a shutter card to block the light coming from the laser while you get your new plate into position.

Just get a small postcard-sized piece of black card; once you have set up your hologram with a "dummy plate," you can then stick a piece of black card into the sand in front of the laser to black the light coming from the laser (see Fig. 8-12). Carefully unwrap your holographic plate from its protective paper, and place it in position where your dummy plate was before. Now remember, every time that we touch the sandbox or move something, we need to allow the "relaxation time" for all the vibration to subside. This being the case, lift the shutter card so it is clear of the sand, as shown in Fig. 8-13. However, ensure that it continues to block the laser beam so that an exposure isn't made. Count in your head the seconds of the relaxation time and then allow a few, or if you

Figure 8-12 *The shutter card inserted into the sand.*

Figure 8-13 *The shutter card lifted out of the sand, but the beam is still blocked.*

have made one, use your audible darkroom timer to mark time. Then when the relaxation time has elapsed, lift the card clear of the beam as shown in Fig. 8-14 and count out your exposure. Once the exposure is made, replace the card in the sand.

Simple Holography

Figure 8-14 *The shutter card is lifted clear of the beam.*

Figure 8-15 *Think creatively about your 3D subject matter.*

Creative Holography

Once you start thinking creatively about what you might like to capture in three dimensions, your family will suddenly find small objects around the house start disappearing as you borrow them for your holography studio to capture them for posterity in three dimensions. Think creatively about how you can use the medium. Create exciting holograms that tell a story by building model scenes to capture, or think of witty jokes that can be captured in a three-dimensional exposure. While walking around a hobby shop, I saw this painter's mannequin, with a miniature easel; my quest was then to find a model of an "artist" that was even smaller. Luckily, a toy packet of merchandise from a well-known children's boy wizard film yielded a model of a mustachioed gentleman yielding a magic wand. It isn't such a great leap of the imagination to see his wand as a paintbrush and the plastic man as an artist with his fine pencil moustache. The visual irony of the model standing behind the easel and painting the artist is amusing as you can see in Fig. 8-15.

Sometimes you might want to support your subject matter in a different way, or you have an awkward object to support. Foam board makes it very easy to rapidly prototype supports for models and subjects (see Fig. 8-16). You can buy

Figure 8-16 *Using black foam board to rapidly construct object supports.*

"black-core" foam board that includes two sheets of black card with a dark foam core sandwiched in between them. This is ideal material because it does not reflect laser light and can be easily assembled with hot melt glue into all manner of complex three-dimensional shapes. It is also firm and rigid, and so will support your materials strongly during exposure.

When it comes to making simple beam reflection holograms, there are fundamental limitations to the depth of image that you can capture, which is known as the *depth of field*. The extent to which you can capture a deep image will depend on what is known as the *coherence length* of your laser. With a good-quality small helium-neon laser, your coherence length is usually in the order of 6 to 12 in/15 to 30 cm. The diode laser from Integraf recommended earlier in this chapter has a coherence length of over a meter when fully warmed up! However, the depth of hologram that you can successfully capture will be around half of

this because the light must make its journey through the emulsion, reflect off the object, and then return to the film. You can see an illustration of this in Fig. 8-17. Remember, the nearer you place your object to the film, the brighter the resulting image will be!

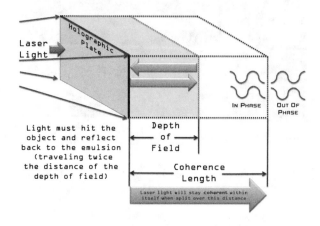

Figure 8-17 *The coherence length of your laser will limit the depth of field you can achieve.*

Project 7: Creating a Single-Beam Transmission Hologram

You Will Need

- Holographic plate
- Laser
- Lens (if using HeNe laser)
- Processing chemicals for hologram

We're now going to create a simple transmission hologram. If you are struggling to remember the difference between transmission and reflection holograms, then just flick back to Figs. 5-21 and 5-22 in Chap. 5 and the associated text for a quick recap. The setup from above that we are going to use is represented in Fig. 8-18.

Using a single beam is very simple; however, it isn't ideal for transmission holograms because the object beam is largely illuminating the objects from the "wrong direction." It's a bit like taking a photograph into the sun; if you're shooting people,

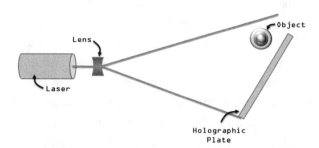

Figure 8-18 *Setup for the single-beam transmission hologram.*

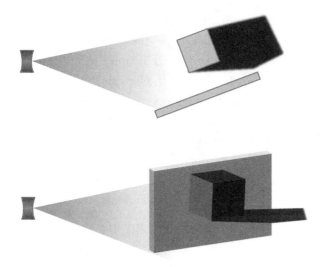

Figure 8-19 *The exposure of a transmission hologram viewed from above and in three dimensions (shadow shown).*

then your subjects will have their faces in shadow while the sun dazzles the film or image sensor! We'll see some more complicated transmission hologram setups in the next few chapters that use multiple beams. However, for now, while we're keeping things simple, we'll make do with the compromises of this method.

You want to make the beam a little more divergent with a reflection hologram. This time we don't just want to illuminate the plate (with the object behind it), but we want to illuminate the object as well.

In Fig. 8-19, we can see the shadow cast by the object, and the two different views, one aerial and another three-dimensional. We've swapped the chess piece for a simple "box" shape because it allows the shadow to be seen more clearly.

Viewing Transmission Holograms

Unlike our reflection holograms that we can view in ordinary light, our transmission holograms are going to require a monochromatic light source to

view them with. This will preferably be the laser that you have created the holograms with in the first place to give the best reproduction. In this respect, diode lasers with the collimating lens are a good bet, for the simple reason that they are lightweight, and much easier to maneuver than a helium-neon laser and diverging lens.

With the collimating lens removed from a laser diode, the output is quite safe to view with your eyes. Of course, don't go looking right into the laser diode up close—treat the beam with respect. But looking at a transmission hologram using the beam once it is spread out is quite acceptable. If you are using a class IIIa diode laser, it will be below 5 mW; once this beam has spread out to 50 mm or more in diameter, it will not be a hazard to the eye.

Hold your hologram in the path of the diverged laser beam at a point where the beam has spread out sufficiently to illuminate the whole holographic plate as shown in Fig. 8-20. Hold the film in the path of the beam, and look into the laser light, not directly at the beam, but through the section of film that is being illuminated. Think of your plate of glass as a window that you are looking through: You should expect to see your object appear on the other side of the glass as if magically floating in space! With a laser diode it is easy to adjust your viewing angle until you start seeing the image. Change your angle in respect to the hologram and the holograms angle in respect to the laser.

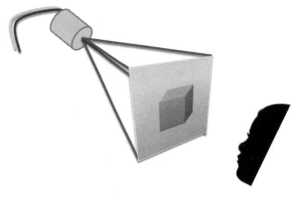

Figure 8-20 *Viewing a simple transmission hologram.*

You should see a glimmer of the image; once you've got this glimmer, finely adjust the angle with which you view the image until you find the point at which the image is brightest. You should find the best image when the beam is in the same position as the original reference beam.

Fun with Transmission Holograms

You might remember that we spoke about every piece of the hologram containing information about the "whole image"; well, there are a couple of ways that we can prove this. The first is destructive. Break or cut your hologram into pieces, and you should see that each piece will still

give you a complete view of the object in question. However, the second method is kinder to your holograms and your wallet! Take a sheet of black card and punch a hole in it. Use this mask as an aperture through which to view your hologram. Hold it right up close to your hologram on the side that you are viewing it, and peep through the hole. As you vary your position, you should see that the whole image can still be seen from the hole—just from this single restricted viewpoint.

The other fascinating property of transmission holograms is that we can "project" them using a collimated laser beam. If you are working with helium neon, then you can use the plain old laser beam without a lens. If you are using a diode laser, you need to replace the collimating lens onto the front of the diode laser, and adjust it so that the laser produces a nice small spot of red laser light.

Project 8: Making a Single-Mirror Transmission Hologram

You Will Need

- Laser
- Lens (if you are using a HeNe laser)
- Piece of black card to act as baffle
- Holographic plate
- Large mirror
- Object

We can improve on the transmission hologram setup while still employing a single beam with the use of a mirror. This helps us to overcome the problem with single-beam transmission holograms of illuminating the rear of the object. Set up your sandbox as shown in Fig. 8-21; you will need a larger piece of mirror than for a lot of the projects that follow because we are bouncing a spread beam, not a fine collimated beam.

By using the mirror, we reflect a reference beam, which hits the plate directly, and also we reflect laser light onto the object, which acts as an object beam. There are some notes in Chap. 9 on beam ratios; if you find that your first result from setting this arrangement up is not successful, take a look at the next chapter, especially Fig. 9-5, and then return to this project.

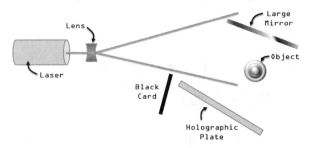

Figure 8-21   *The single-beam transmission hologram with mirror.*

Project 9: Creating a Multiple-Channel Hologram

You Will Need

- Holographic plate
- Two different objects
- Laser
- Lens (if using HeNe laser)

If you thought that the amount of information that we can store using a hologram (all the simultaneous views of a 3D scene) is incredible, this project will make you think yet again about holography! If you skip ahead to the chapter on the future of holography; you will see how holographic versatile discs have the potential to radically increase the amount of information that we can store on a disc the size of a standard DVD/CD. How will they do it? In this project, we're going to show that by creating a multiple-channel hologram, we can store not just one, but two scenes on a single holographic plate, just by changing the incident angle of the reference beam. You can use this to dramatic effect by careful selection and positioning of objects. I've used chess pieces so far in these diagrams, so for this exercise, let's pick a pawn and a bishop. It doesn't really matter what objects you use; however, you want to consider picking something small that doesn't fill the entire frame too much to ensure there is not too much interference between the two images that you are trying to capture.

We're going to make two different exposures this time. We can see the setups for these two exposures in Fig. 8-22. One of the things you will

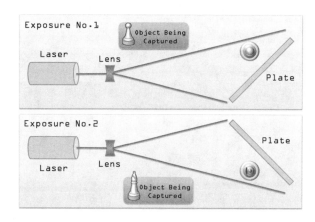

Figure 8-22 *The two exposures for a multiple-channel hologram.*

have to remember is that we don't want to overexpose the plate. If we are making two exposures, we need to make each exposure for half of the time.

In between the exposures, when the laser light is blocked, we're going to change the object, place it in a "mirror image fashion," and rotate the plate through 90°. This causes two separate images to be exposed on the same plate. When we develop this plate, what we will find is that two different holograms can be viewed from two different angles.

In fact, once you've proved you can do this with two exposures, there is nothing to stop you from making more exposures with additional angles and objects. Just remember, that each time you add an exposure, you need to reduce the time for that exposure. The sum of all the individual exposures should be the same as that which you would give a holographic plate if you were making a single exposure.

Chapter 9

Intermediate Holography

In this chapter, we turn up the heat a little (although not during an exposure obviously because that could ruin our interference pattern) and start working with more complicated holography. For a start; we're going to look at using a film. Any of the projects in this book will work with film. Where it says "plate," read "film," and vice versa. They are interchangeable except to begin with using plates is recommended because film can be a bit fussy to work with for beginners. Check out the first part of this chapter for all you need to know.

Then we're going to look at some slightly more complicated holography setups that employ beam splitters to work with two beams. We'll save all the mad crazy stuff for the advanced chapter, but here if you have successfully worked through the basic chapter, you should find more challenge and intrigue.

Project 10: Working with Film

You Will Need

- Glass (antireflection coated glass much preferred)
- Springback or bulldog clips
- Black electrician's tape
- Index matching fluid

Working with film is a little trickier than working with glass plates. Because film is coated onto a flexible base, it can move easily, which can result in ruined holograms!

If you plan to use film, you need to prove some mechanical support to stop the film from moving around. There are a couple of simple ways to do this:

You can use two pieces of glass to sandwich the film in between as in Fig. 9-1, and then clamp this

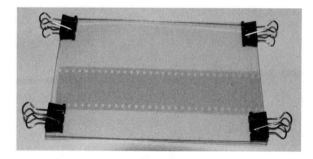

Figure 9-1 *Flexible film clamped between two rigid glass plates.*

sandwich together with a pair of sturdy clips. Apply gentle pressure to this sandwich to ensure that all of the bubbles of air that could become trapped between the film are removed from the sandwich.

Another alternative is to make a foldable glass sandwich, with some black electrician's tape. Two sheets of glass are joined by a "hinge" made of electrician's tape. Within the sandwich, you can provide an "L" of electrician's tape (carefully cut to avoid doubling up the thickness). This will help when you want to align your film in the sandwich. Place two tabs of tape at the top. You can see this sandwich open in Fig. 9-2. Fold the sandwich and use the tabs to secure it, smoothing out any air bubbles as you can see in Fig. 9-3.

If you develop your film holograms and find a series of fine lines all over your holograms, you are suffering from the curse of Newton's fringes. Although admittedly Newton had a decidedly dubious hairdresser, it's nothing to do with his tousled tresses; it is more to do with the reflection of light between two surfaces. We've seen all about interference already, so the concept should be familiar to you. If you want to get rid of them, start using antireflection coated glass, or alternatively, you can use an "index matching fluid," which is something that will fill the void between film and glass to prevent reflections. In a pinch, you can use ordinary household white spirit. Because this liquid has a high surface tension, you can get away with a single sheet of glass. The white spirit will make it cling to the glass firmly.

Figure 9-2   *A film holder made from two sheets of glass and some black electricians tape (opened for film insertion).*

Once you've made your glass-film sandwich, you can treat this as if you were using a holographic plate. You might want to revisit some of the simple holography projects, but this time using film; or if you are confident you might want to work your way through some of the slightly more complicated projects in this chapter.

4 × 5-in sheets of film carry a notch on one corner that can be used to identify the emulsion side of the film, as seen in Fig. 9-4.

Figure 9-3   *A film holder made from two sheets of glass and some black electricians tape (closed).*

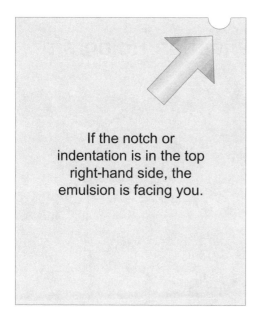

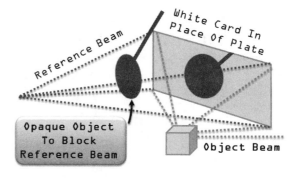

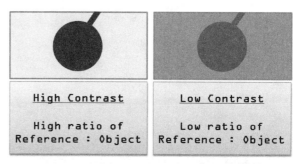

If the notch or indentation is in the top right-hand side, the emulsion is facing you.

Figure 9-4 *The notch in sheets of flexible film can be used to identify the emulsion side.*

Type of Hologram	Reference Beam		Object Beam
Split Beam Transmission	4	:	1
Split Beam Reflection	2	:	1
Reflection Image Plane	1 ½	:	1

Figure 9-5 *Gauging the relative strength of object and reference beams.*

Creating Split-Beam Holograms

In this section, we're going to start using split-beam setups. This will introduce the beam splitter to create separate object and reference beams. There are a couple of ways that we can gauge the relative intensity of the different beams. The first is to use our photometer, discussed a little more in the "Your Holography Workshop" chapter. With this you will be able to take readings of each individual beam. You will need to ensure that your meter is set to the correct scale readings in order to give a meaningful output. This is a highly accurate way of doing things; however, there is a simpler method. Use a piece of card on a stick (as shown in Fig. 9-5), or even your thumb to block the reference beam, and see how it shadows the film plate. By looking at the contrast between the reference beam and object beam, we can often make a simple assessment of the nature of

the ratio between the reference and object beams.

There are couple of golden rules to be followed with split-beam holography that are worth bearing in mind. Don't pass a laser beam over the laser itself. Heat from the laser has the potential to make things distorted. Furthermore, you may find that you end up with nuisance reflections from the edges of optics, reflections of optical stands, and the like. The way to get round this is to paint everything that isn't an optical element matte black (i.e., stands, supports, etc.), and use pieces of black card poked into your sandbox to mask off any unwanted extraneous light.

Project 11: Multiple-Beam Reflection Hologram

You Will Need

- Laser
- Beam splitter
- 2x mirror
- 2x lens
- Plate
- Object
- Holography-processing chemicals
- String

By creating a separate beam, we can make a multiple-beam reflection hologram where we have more control over the object beam. This setup is illustrated in Fig. 9-6. This has advantages over a single-beam hologram. We still have the advantage of reflection holograms of being able to view them with white light.

When you are setting this up, use a piece of string to measure the path lengths of the laser light

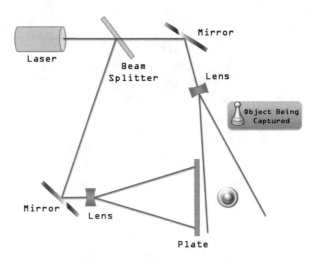

Figure 9-6   *A split-beam reflection hologram.*

from the beam splitter. We want to try and make the object and reference beams as equal as possible. We want to try and get beams of equal strength reaching the plate.

Remember to keep the emulsion side of the plate or film facing the object and to locate the object as near to the film as possible.

Project 12: Split-Beam Transmission Hologram (I)

You Will Need

- Laser
- Beam splitter
- 2x lens
- Mirror
- Large mirror

- Holographic plate
- Object

Figure 9-7 shows a split-beam transmission hologram setup. As with the reflection hologram that precedes this project, we still want to try to keep the lengths of the optical paths of the object beam and reference beam as similar as possible for best results. The trusty piece of string will come in handy here, allowing you to make adjustments as and when necessary.

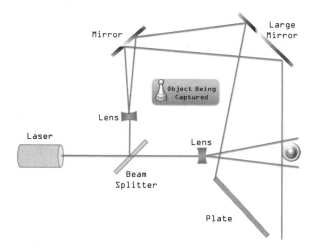

Figure 9-7 *A split-beam transmission hologram setup.*

For small objects, 3:1 or 4:1 will work well; however, as your object starts to fill more and more of the frame, you might want to water

down the object beam a little to ratios as little as 10:1. If this is hard to achieve with the lens, you can purchase neutral density filters. The analogy is a little bit like a resister in an electrical circuit. By inserting a neutral density filter, you can resist the flow of light and make the object beam dimmer.

One thing to watch for with this setup is that because the reference beam is spread wide early on, we need to ensure it is sufficiently bright (or if you prefer to look at it another way, that our object beam is sufficiently dim!) to get the right beam ratio. Play around and use an N.D. filter if necessary.

The bigger plates you expose and the more you spread your beams out, the longer your exposure times are going to be. Be warned because this makes you vulnerable to vibration.

Project 13: Split-Beam Transmission Hologram (II)

You Will Need

- Laser
- 2x mirrors
- Beam splitter
- 2x lens
- Holographic plate
- Object

Figure 9-8 shows an alternative split-beam transmission hologram setup. You may need to adjust the "beam ratios" of the object and reference beam. Until our laser beam hits the diverging lens, it is concentrated into a narrow beam of light; however, once it hits the diverging lens, inverse square law kicks in, and as the beam

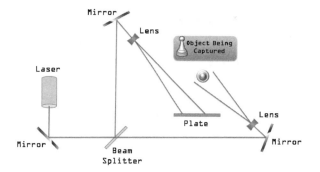

Figure 9-8 *Another split-beam transmission hologram setup.*

spreads out more, it decreases in intensity. This gives us some room for adjustment. By moving the diverging lenses along the line made by the laser beam, we can adjust the ratio of the reference beam to object beam to as near to the optimum of between 3:1 and 4:1 as possible. Our light meter and/or shadow method should come in handy here!

Project 14: Transmission Hologram with Soft Lighting

Project 14: Transmission Hologram with Soft Lighting

You Will Need

- Laser
- 3x mirrors
- 2x lens
- Holographic plate
- Object
- Sheet of opal/frosted glass
- Beam splitter

Direct illumination with laser beam is all well and good, but it can give rise to some rather bleak shadows, and sometimes this isn't the creative effect that we desire. It's worth knowing how we can manipulate our laser beams to create different visual effects. Remember, holography is at the juncture of art and science, and sometimes we need to employ a bit of science to create an artistic effect.

By sticking a bit of frosted or opal glass in between our reference beam and our object, we can create a softer effect; as can be seen in Fig. 9-9.

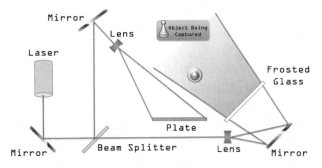

Figure 9-9 *A transmission hologram with diffuse lighting.*

Taking Split-Beam Setups Further

All of the setups in this section have used simple, cheap, and readily available components; however, once you get confident with split-beam setups, there is plenty of room for experimentation and trying things differently. This will of course depend on what equipment is available to you and what budget you have.

You could consider replacing the diverging lenses in this setup with small spherical mirrors. Although this will change the geometry of the setups, the result will be similar.

Furthermore, there are a number of things we can do to improve the quality of your images. For a start, if you use a laser with a polarized output, this will help greatly. We've not delved into the use of half-wave plates here because the projects have been selected for beginners; however; if you read some more advanced holography texts, you will see how these can be employed to improve image quality.

In the optics chapter, we briefly touched on the idea of "collimated beams" and different optical arrangements we could use to create these. Substituting a simple spread beam from a lens for a collimated beam is another improvement you can make.

Finally, if your wallet (or a kindly relative's) will stretch to a pair of spatial filters, you can make vastly improved holograms due to the clean nature of light from a spatial filter.

If you have managed to get some of these projects to work successfully, get prepared to enter the mad, crazy world of advanced holography.

Chapter 10

Advanced Holographic Projects

Project 15: Making a Hologram with Diffuse Illumination

You Will Need

- Laser
- 3x mirrors
- Beam splitter
- 2x lens
- Object
- Diffuser (see later)

So far, even in the split-beam setups that we looked at in the intermediate chapter, the light from our laser is still quite stark. There are occasions, when if we are using holography creatively, we do not want such a contrasting source of illumination for our object. If this is the case, we can use a frosted diffuser to make our "object beam" more diffuse. Think about the different quality of light emitted by an incandescent light bulb with a clear bulb, compared to, say, a frosted bulb. The light from

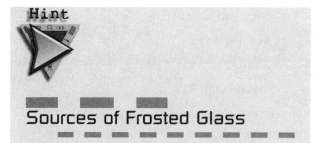
Hint

Sources of Frosted Glass

We need to create a diffuser to take the laser beam and make it softer so that our object beam does not cast harsh shadows. You have a few options as to how to create this diffuse light; ideally you want to source a sheet of opal or finely frosted glass that you should be able to get from your optics supplier. At a pinch you could use a sheet of greaseproof or tracing paper, but ideally you want a piece of fine frosted glass. You can buy "glass frosting spray" from housewares stores as an alternative.

the clear bulb is very stark and casts dark shadows, whereas the light from the frosted bulb is a little softer and more diffuse.

Figure 10-1 shows the setup for making a hologram with diffuse illumination.

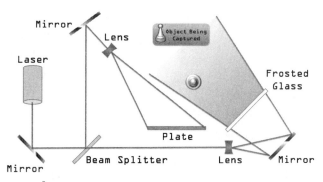

Figure 10-1 *Making a hologram with diffuse illumination.*

Project 16: Making a Hologram with Multiple Sources of Illumination

You Will Need

- Laser
- 4x front surface mirrors
- 2x beam splitters
- 3x lenses
- Holographic plate

Sometimes we want to make a hologram that is lit from multiple directions. It is possible to create multiple object beams while using only one reference beam; it requires another beam splitter mirror and lens, but it can be used to soften shadows and make some subjects appear more pleasing. Figure 10-2

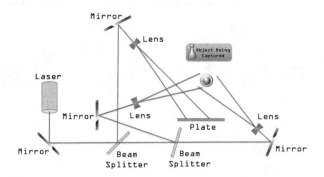

Figure 10-2 *Making a hologram with multiple sources of illumination.*

illustrates an example of this type of setup. Remember that you want to keep all beam paths as equal as possible, and that for a transmission hologram of this nature, you want an object to reference beam ratio of between 3 and 4 to 1.

Project 17: Making a Copy of a Hologram

You Will Need

- Index matching fluid
- Triethanolamine
- Glass plates
- Laser
- Diverging lens
- Master hologram

One of the methods that we can use to copy holograms is to "contact copy" them; this is where the master and transfer films are kept in close contact and an exposure is made. Index matching fluid must be used between the films (an explanation of which is given in Chap. 9).

Copying Reflection Holograms

A master reflection hologram needs to replay in the right color for a strong image to be produced, using a very small angle of light or swelling the master hologram using triethanolamine (see Chap. 11) can be used to ensure that the hologram replays in red to allow a red laser to be used for the transfer. Remember, if you are making a transfer of a reflection hologram, the laser light will need to pass through the transfer film before it passes through the master hologram. Furthermore, we can adjust the position of the copy hologram plate to intercept the real image or place it in front of or behind the real image to change the position

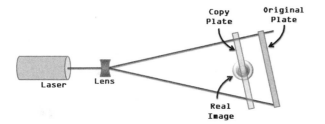

Figure 10-3 *The position of the copy plate can be adjusted to coincide with the real image.*

of the image as it is represented on the copy hologram. Figure 10-3 illustrates this.

Copying Transmission Holograms

With a transmission hologram, you must first use a laser to find the angle at which the hologram replays the brightest possible image. When making a transmission hologram, the copy will need to be located on the side furthest from the laser. You can then place the plate to intercept this image, with another beam set up as a reference beam.

Project 18: Experimenting with 360° Holograms

You Will Need

- VW Buggy 360° hologram (source from Integraf Item # H-VWBUG)
- Laser
- Lens (if using HeNe laser)

In Chap. 11, we're going to start looking at holograms that rather than just using a plain flat plate use holographic media creatively to not only produce a three-dimensional (3D) view of the object in question but produce a 3D view that encompasses all angles of the object under examination.

Before making your own 360° holograms, it's nice to take a look at a professionally made hologram. Integraf sells a small 360° cylindrical hologram of a Volkswagen Beetle, which is fascinating to behold. Although the pictures do not do this hologram justice, Figs. 10-4 through 10-6 show the VW Beetle hologram from a variety of different perspectives.

To view 360° holograms, you have a number of options (see Fig. 10-7). You can hold a sheet of frosted glass on top of the cylindrical hologram

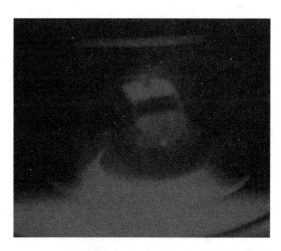

Figure 10-4 *VW Beetle hologram viewed from the front.*

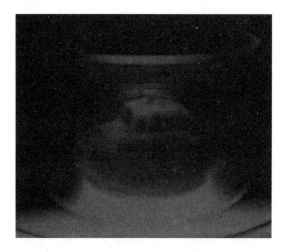

Figure 10-5 *VW Beetle hologram viewed from the side.*

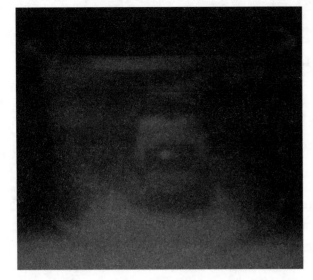

Figure 10-6 *VW Beetle hologram viewed from the rear.*

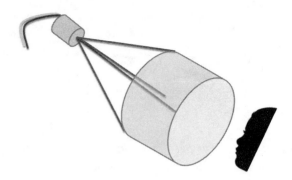

Figure 10-7 *Viewing 360° hologram.*

and shine a laser onto this to scatter the light; however, it is better to use a spread beam that shines to illuminate the whole cylinder. Then look through the sides of the cylinder and into the cylinder to see the different effects that are created.

Project 19: Making a Direct-Beam 360° Cylindrical Hologram

You Will Need

- Short length of clear plastic pipe of diameter large enough to encompass your object

  Or

- Straight-sided jar

  Or

- Industrial Fiber Optics Holography Kit

- Film (cut to the dimensions to form a cylinder within the length of the pipe/jar)

- Object to holograph

- Laser (diverging lens or lenses if you are using a HeNe laser)

This is a holography setup that looks deceptively simple; however, due to the tricky nature of positioning everything, it can be fiendish to execute well. Cutting to the quick, we are going to surround an object with a cylinder of holographic film (see Fig. 10-8). We need to support this film inside a clear

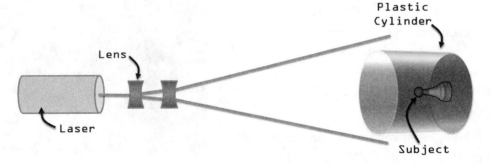

Figure 10-8 *Setup of direct-beam cylindrical hologram.*

Figure 10-9   *Picture of direct-beam cylindrical hologram.*

cylinder, which can be an old jar, or plastic container or piece of acrylic tubing. Alternatively, if you purchase the "industrial fiber optics" kit mentioned in the You Will Need list, you will find a perfectly shaped plastic cylinder designed for this project.

You have a couple of options here, depending on your budget and outlook. Industrial Fiber Optics, listed in the suppliers' index, has prepared a plastic jar, lid baffle, and convex mirror that are absolutely perfect for this experiment, if you are prepared to buy their holography kit. This is a good investment because it comes with a handful of different optics, some film, and chemicals.

Of course you need to secure your object horizontally; use a small plinth to raise your object from the bottom of the jar. This could be a small lid from a bottle, for example. Furthermore, you will want to ensure that the object and whatever you are standing it on are all mounted rigidly together. A small dab of hot melt glue will hold everything together tightly, and with a little gentle force should be easy to remove without causing damage to the base of the object, the stand, or the cylinder if applied carefully. See Fig. 10-9.

If you are working with a helium-neon laser, you will most likely find that a single lens is insufficient to spread the beam enough to ensure that the lip of the jar is illuminated all around. If this is the case, you will need to use a pair of lenses to increase the spread of the beam. This is illustrated in Fig. 10-10.

One of the problems with performing this setup in a sandbox is that sand reflects light; you don't want light being reflected from the sand onto the sides of your cylinder because it has the potential to

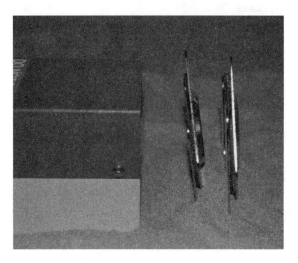

Figure 10-10   *Close-up of laser and double lens arrangement to spread beam.*

compromise the quality of your hologram. Therefore, support the cylinder in the sand, on a sheet of bent black card or foam board to ensure that the only light the film receives is from inside the jar. This can be seen clearly in Fig. 10-11. You may also want to cut a

Figure 10-11   *Ensure the cylinder is supported by a black surface so as not to reflect light.*

circle of black paper to place in the base of the cylinder if you are using a jar, to prevent unwanted reflections from the bottom surface.

Exposure Ready Reckoner

A photometer is of little use when it comes to this setup. Instead, use this ready reckoner for guidance.

Laser Output Power	Exposure Time
5 mw	Half – 1 s
3 mw	1–2 s
1 mw	3–4 s

This is quite a complicated setup, and unlike other holograms where we have the film sandwiched between two plates, here the film only has support from one side, so allow a few minutes of relaxation time.

Processing Your 360° Hologram

Now you will be processing a strip of film that is larger than many of the plates and film that you have been working with up until now, so you will need to consider finding an appropriate size container that will allow the whole strip of holographic film to be immersed in the chemical solution in order to ensure an even development across the whole area of the film.

Viewing Your 360° Holograms

Follow the guidance in the previous project when it comes to viewing your hologram. However, bear in mind that because your hologram was produced within the plastic cylinder, it should be placed back there for viewing purposes to help the film retain its structure and rigidity. Remember, once you've processed your film, you want to stick it back so that the emulsion continues to face inward. If you can't see an image or you can see colors but it appears unclear, then make sure that your film hasn't been inserted into the cylinder upside down or inside out. You want to try and recreate the conditions under which the hologram was created, so if you used a convex mirror in the bottom of the jar, ensure that it is there again.

Project 20: Making a Cylindrical Hologram with a Convex Mirror

You Will Need

- Components as above
- Baffle (see later)
- Convex mirror (see later)

Hint

You need a baffle that will stop the laser light from entering the side of the cylinder or the walls of the jar that you are using and will also prevent the divergent beam from hitting the film directly. You can use a compass cutter and a sheet of foam board to produce a "donut"-shaped piece of foam board that when affixed with Blu-Tack will suffice. Industrial Fiber Optics sells a great holography kit that comes with a jar with a screw-on plastic black rim especially for this purpose. The kit is reasonably priced and comes highly recommended. If you are going to make your own, there are many expensive circle cutters on the market sold by companies that supply the professional picture framing trade; however, there is a cheap economical model sold by Olfa that is perfectly adequate for the job. It is available from a number of online retailers. Search for "Olfa Compass Cutter."

Hint

Again, Industrial Fiber Optics sell a kit with a convex mirror designed to fit into the jar supplied with the kit. If you are struggling to find a mirror that will fit in your cylinder; try finding a silver plastic bauble Christmas decoration. Although it won't make a top-quality optic; a slice cut from a large (plastic!) bauble should more than suffice.

Figure 10-12 illustrates the cylindrical hologram setup. You will see the mirror positioned in the bottom of the jar. Light that doesn't hit the object directly will be reflected off the mirror and create the reference beam on the film. We don't want the divergent beam hitting the film directly this time.

Figure 10-13 *Notice the black baffle and convex mirror.*

You can see the jar close up in Fig. 10-13. If you want to make this hologram horizontally, you will need Blu-Tack to secure the mirror in the bottom of the jar. Figure 10-14 shows a photograph of the entire setup. Although Fig. 10-15 shows a close-up of the jar in the final setup, remember you can use black card to prevent reflections from the sand.

You may find, depending on the equipment you use, that you have trouble securing the mirror in the bottom of the cylinder. If this is the case, you can angle your laser and lens arrangement vertically and then "bounce" it down with the help of a mirror. This is all illustrated in Fig. 10-16 as a diagram and in Fig. 10-17 as a photograph.

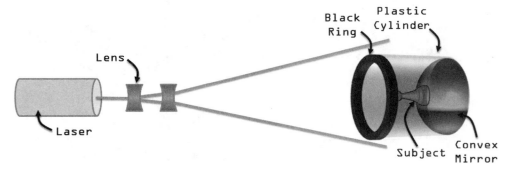

Figure 10-12 *Setup of cylindrical hologram with mirror.*

Figure 10-14   *Photograph of the setup.*

Figure 10-15   *Photograph of the close-up of the jar, mirror, and cylinder with the object stuck to the mirror.*

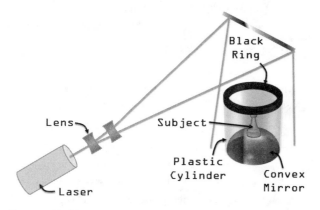

Figure 10-16   *Setup of the alternative exposure method.*

Figure 10-17   *Photograph of the alternative exposure method.*

Project 21: Making a Conical Hologram

You Will Need

- 4 × 5-in sheet of holographic film
- White foam board
- Your regular hologram-making equipment

Tools

- Hole punch
- "Circle cutter"

We can do something similar to our holographic cylinders that we have just created, albeit with a cone of holographic material. This has some advantages. By forming the film into a cone shape, we give it some rigidity, so there is no need for the additional "jar" to support the film.

Use the exposure setup shown in Fig. 10-16, but substitute the cylinder for a cone of film, as shown in Fig. 10-18.

Making a conical hologram really is as simple as cutting a cone of holographic material, joining it to form the shape, placing it over an object, and exposing it to laser light from above. Conical holograms are only ever going to be reflection holograms because it is near impossible to make a transmission hologram due to the need to get the beam "into the cone."

To make the film bend easily, use a hole punch to remove the film at the center of the cone, which will allow you to curl it easily. Cutting out a cone by safelight really is the hardest bit! Make sure when you cut that you remember what side of the cone has the emulsion on because when you curl it, you will want to make sure this side faces into the cone. Use a bit of cellophane tape to secure your cone, and put the join at the "back" of your subject because this area will not capture well.

Figure 10-19 shows a cutting diagram that will produce a cone from a 4 × 5-in sheet of film, and

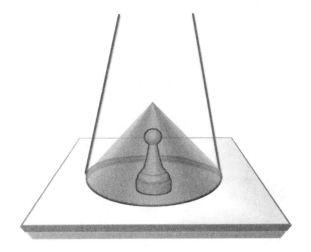

Figure 10-18 *Setup for the conical hologram exposure.*

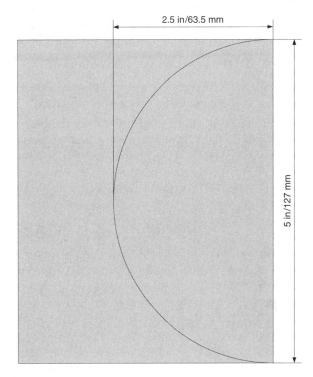

Figure 10-19 *Cutting diagram to produce a cone from a 4 × 5-in piece of film.*

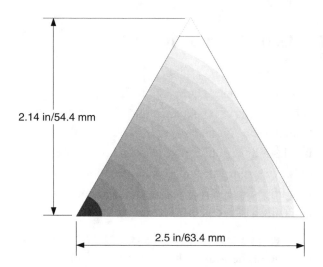

2.14 in/54.4 mm

2.5 in/63.4 mm

Figure 10-20 *Volume of a cone produced from film cut to Fig. 10-19 specifications.*

Fig. 10-20 shows the volume of the cone that you must enclose your object within.

If you're a frugal holographer, you'll hang on to the scraps of film and put them in a lightproof envelope to use later for test exposures.

To make a support for our cone, take a piece of white foam board and a circle cutter. A circle cutter looks much like pair of compasses; however, instead of having a pencil, you will have a scalpel blade. Be exceptionally careful because this blade is *sharp!* Set the cutter to half (the radius) of the diameter of the base of your film cone, and placing the point in the center of a piece of foam board, cut a circle. Remove the circle. We want the piece of board that you have cut it from, rather than the circle itself. Stick this board on top of another piece of foam board using double-sided tape. You should now have a nice recessed circle in which your hologram cone will fit. You can also use a piece of Blu-Tack or glue to secure the base of your subject to the base piece of foam board. Place the cone on top, and use a little glue or Blu-Tack to affix the edges to the circle support.

You will want to set up your laser so the spread beam shines on top of the cone. Alternatively, glue some supports to the foam board, and drive it into your sand, so that the cone points toward the spread laser beam horizontally. Once you've completed the film/cone setup, leave this on your sandbox to settle for quarter of an hour.

Make the exposure, and then disassemble the cone carefully, making sure not to scratch the emulsion. Process the film, and when you are done, reassemble with another little strip of tape. Illuminate from above with a spotlight!

Here's an idea; one which I'm yet to try; but one for someone with sufficient patience and resources. Make a conical hologram chess set! Chess pieces are popular holography subjects, so make holographic replicas of all the chess pieces, and mount them to foam-board bases. Black pieces wouldn't make good holograms, so make two sets of holograms of white pieces! You could identify your pieces, and mark them with bands of black and white tape where the hologram joins to the foam-board base. You'd have to play under a white spotlight (which would create a certain tense atmosphere)!

Project 22: Making a Hologram Cube

You Will Need

- Four holographic plates
- Your regular hologram-making equipment

Tools

- Glue gun

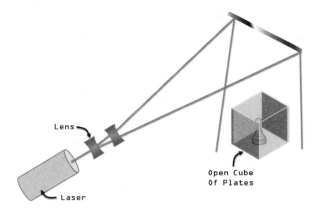

Figure 10-21 *Setup for hologram cube.*

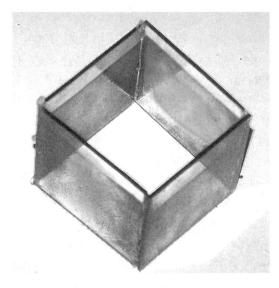

Figure 10-22 *Hologram cube made from glass plates joined with hot melt glue.*

Another variation on a similar theme is to replace the film cone with a cube made from plates as shown in Fig. 10-21.

You can take four plates, and with a little bit of careful fun with a glue gun (in the dark or safelight, so be careful of burns!) you can join them to form a cube as shown in Fig. 10-22.

You can then expose this as you would with a cylindrical hologram from above. The beauty of hot melt glue is that you can peel it off relatively easy when you've made your exposure, ready for you to process it. A cube hologram is quite useful if you want to make a hologram of something with four distinct sides—for example, a model

house or car. For circular objects, the cylindrical or conical holograms give a better effect!

Any 3D shape can effectively be made if you have the patience to sit and cut out the pieces from glass plates. The only limitation is the weight of the plates and how well you can support them. Pyramids are also easily possible, as are octagonal prisms. Beyond this you'll need deep pockets and to be a crack shot with the glass cutter, but anything is possible if you have the patience!

Project 23: Rainbow Transfer Hologram

You will recall from our discussion of the history of holography that Dr. Stephen Benton discovered the principle of the "rainbow" hologram (see Fig. 10-23); from which we can then produce the embossed Mylar holograms that are so common in everything from credit cards to authentication on designer goods.

If you think back to the chapter on 3D vision, we see in three dimensions because our eyes capture two images in the horizontal plane. So to satisfy our requirement for "3D-ness" (not an

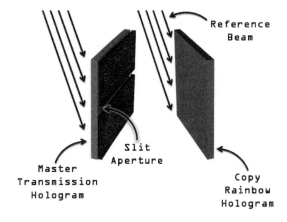

Figure 10-23 *Rainbox hologram setup.*

official word), we need information about different horizontal perspectives, but really the vertical information is to some degree superfluous. We would lose vertical parallax when we change our vertical position in respect to the hologram, but the key horizontal parallax is still there.

We can make a transfer hologram but remove all of the vertical information by copying the hologram through a "horizontal slit." This removes all of the vertical information. In its place, when we view the hologram using white light, all we see is a change of color, giving rainbow holograms their characteristic appearance (shown in Fig. 10-24).

There is another alternative. If you can get either a cylinder of glass (or even a test tube filled with glycerol), you can make a lens that only works in one dimension! This will expand the beam in a "slit" and enable you to copy without masking.

Interestingly, take a rain box hologram and look at it using laser light, and you will see that the image can only be seen through a "slit"; it's a bit like peeping through the letterbox. Move up and down. and the image disappears. This is because the rainbow hologram is diffracted only at a certain angle in respect to the horizontal. When we shine white light through the hologram, different wavelengths are diffracted at different angles.

You can also create a focused image rainbow copy with a lens and slit as shown in Fig. 10-25.

Points for experimentation:

- See how changing the size of the "slit" affects the sharpness of your image.

- See how changing the distance between the master plate and copy plate changes the position of the hologram in the copy.

- See how changing the reference beam changes the position of the colors displayed with respect to the angle in which you view your hologram.

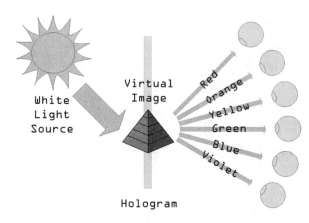

Figure 10-24 *A rainbow hologram's characteristic appearance.*

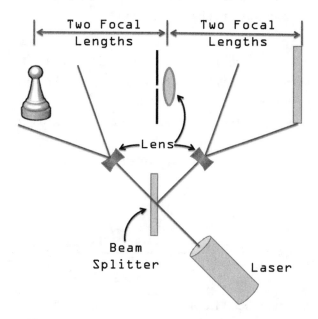

Figure 10-25 *A focused image rainbow hologram copy.*

Chapter 11

Advanced Holographic Chemistry*

Understanding the Chemical Processes That Produce Holograms

In the emulsion of your hologram, there are light-sensitive chemicals known as *silver halides* that react to the presence light of certain wavelengths.

You can think of the emulsion as a lot of very small crystals called microcrystals, dispersed throughout the gelatin emulsion. It's a bit like if you made some gelatin dessert, but as it was setting, you mixed in a bucket of sand. It would taste awful, but it helps you to visualize on a bigger scale what is happening on the microscale of a photographic emulsion.

We take a plate, which is "unexposed," and on the microscale, some of this plate will be exposed to light while other portions of the plate will not. At the "macro" scale in which we view the plate, it appears that all of the plate is receiving red light; however, if we were able to zoom in to the nanoscale on which wavelengths are measured, we would see patterns of light and dark where reference and object beams interfere. This is all discussed in detail in Chap. 5.

The gelatin from which the emulsion is made does several jobs. It supports the crystals of silver in place, it allows light to pass through, but it also allows the chemicals in the developer to reach the crystals of silver. When we immerse the plate in developer, the gelatin swells and expands slightly as it moistens from the developer—and the process of development occurs. We then bleach the image,

and the areas that were not exposed to light are washed away. After bleaching, we dry the hologram, and the gelatin shrinks back to its original state as the water is released. This is illustrated in Fig. 11-1.

But how do the light-sensitive silver crystals work? The chemicals are known as silver halides because they are a compound of metallic silver (as

Unexposed

Exposed

Development

Bleaching

Drying

Figure 11-1 *The process of exposure and development.*

---

*Grateful thanks are expressed to Jeff Blyth.

you'd find in jewelry), and one of the chemical elements that is referred to in the series known as "halogens." The halogens are found in group VII in the periodic table, or if you are using the International Union of Pure and Applied Chemistry (IUPAC) style, group 17. In nature, because halogens are very reactive, we find them as compounds or ions.

Two of the halogens are of no interest to us. Fluoride is soluble in water, and astatine is highly radioactive. Therefore in holographic emulsions we focus on using silver chloride (from chlorine), silver bromide (from bromine), and silver iodide (from iodine).

Most photographic emulsions are sensitive to blue light, but not to red, which is why in conventional black-and-white photography, we often use a red safelight. The red-sensitive emulsions that we use with our helium-neon and diode lasers are sensitized to short wavelengths, however, with the addition of other special sensitizing dyes. We can make the emulsion sensitive to different colors. This is why if you want to start using green lasers, you're going to need to use different materials that are sensitive to green light!

With our holographic emulsion, light, from the object and reference beams, falls on the surface of the emulsion. The silver halide particles absorb the energy of the incoming light, as shown in Fig. 11-2.

This light energy helps to break some of the bonds in the crystalline structure of the silver halide, resulting in silver atoms, which are contained within the crystal being released. Once enough silver has been released, a speck of metallic silver forms. This is what is called the *latent image*, which the process of development turns into a visible image.

The process through which this latent images is turned into a real one is known as *chemical development*. This happens by a process of reduction; whereby the silver halide crystals "gain electrons." An easy mnemonic to help you remember whether electrons are being lost or gained in reduction or oxidation is "LEO the Lion goes GER": Loss equals oxidation, gain equals reduction. Memorizing Fig. 11-3 should help you remember this.

Once a grain of silver has formed, it acts as a microelectrode, bringing the developer into electrical contact with the silver ions of the emulsion; the process of development intensifies this "latent" image into dark metallic silver.

In conventional photography, we're familiar with the idea of "negatives": White areas appear

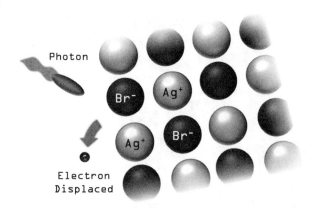

Figure 11-2 *Silver halide particles absorbing the energy of incoming light.*

Figure 11-3 *A simple way to remember the relationship between electrons; oxidation, and reduction.*

black and vice versa. However, in holography, remember we are recording interference patterns, not absolute intensity, so the concept of "negatives" is not applicable.

Exposure Time

If you've ever used a cheap disposable camera; you'll understand how tricky it is to expose film correctly. In picking an aperture and shutter speed, the manufacturers of these cheap plastic marvels have to strike a compromise between the whole range of conditions that the photographer is likely to anticipate. And despite the tricks that photo labs can use to enhance the pictures, the chances are that there are going to be a few pictures taken in low light that look horribly dark and grainy, whereas there will be a few pictures taken in the sun that look bleached out and too light.

As with this photography example, holographic emulsions are equally intolerant of poor exposure; being a chemical process, we need to ensure that the right amount of light reaches our film to create the exposure; too little and the image will be weak, too much and the image will be bleached out. Either way you can't win, so the trick is to find the sweet spot right in the middle!

We cover in this book how to build a simple photometer, which you can use to help you make judgments of how much exposure you need to give your film for any given setup; but taking a brief look at the math that underpin holographic exposure chemistry will aid you immensely in making sure that your holograms get the optimum amount of laser light.

Any given holographic emulsion will have a given sensitivity; if you can remember again far enough back to the days of film cameras, you'll remember that on bright sunny days, you used an ISO 100 film, whereas if you wanted to make short exposures for capturing fast action or if you were

shooting in darker conditions, you'd need a more sensitive film, and the ISO number would increase. Well, holographic emulsions are a little like this; only you are relatively limited in your choice of sensitivities, so you work with what you have by adjusting the exposure time. Besides, on a holographic bench, this is little hassle. We're not capturing motion, so it's just a case of making a longer or shorter exposure.

The manufacturer of a holographic emulsion will express the sensitivity of its emulsion using the units "joules per square meter." From physics class, "joules" should seem familiar as a measure of energy, whereas "square meter" clearly indicates an area. So it should seem apparent that what we are measuring is the amount of "energy" a holographic emulsion needs to receive over a given area to correctly expose the film, which is to say neither "overexposed" nor "underexposed." It should also be apparent that the type of energy we are using to make our exposure is "light energy." However, it's not quite as simple as that. When we say light energy, we don't mean just any old light energy. We mean light energy of the wavelength that the film is designed to be sensitive to. The holographic films we are working with in these projects are designed to be "red sensitive," so the Slavich PFG-03M plates we use are sensitive through the range of wavelengths 630 to 660 nm; the Slavich PFG-01 film is sensitive through the range of 600 to 680 nm. A quick browse of a holography supplier's Web site should reveal a plethora of other emulsions, all with different sensitivity characteristics.

However, back to our measurement of sensitivity, we know the sensitivity is expressed as a unit of energy over an area. Energy is power over a length of time, so our "joule" measurement of energy is in fact equivalent to "power over time." One joule is equivalent to the power of one watt, over the period of one second.

From our knowledge of lasers, we should have picked up now that their power is expressed in "milliwatts," fractions of a watt. It should start to become apparent, therefore, that if we know the "power" of a light source reaching a holographic plate, and we know how much "energy" the plate needs to receive for a correct exposure, we should be able to compute the "time" that power of light needs to shine to give the plate the cumulative amount of energy to make a correct exposure.

We can use the following formula shown in Eq. 11-1:

Equation 11-1: Exposure Formula

$$\frac{\text{Required Energy (Joules)}}{\text{Power of Source (Watts)}} = \text{Time (Seconds)}$$

Now this isn't an exact science because there is some variation in the manufacturing of holographic emulsions, and you may find that one batch of emulsion has a slightly different characteristic from another. This makes things particularly difficult, and if anything only builds the case for buying large boxes of plates so that you get a run of exposures using very similar materials.

If we have an accurate photometer (there is a circuit in Chap. 13 that we could calibrate against a known light source or a commercial photometer) and we know the area of the photo sensor used by the photometer, then we can work out the exposure time.

Take a look at Eq. 2: Calculating the power density on the photometer cell.

$$\frac{80 \ \mu\text{W}}{2 \ \text{cm}^2} = 40 \ \mu\text{W/cm}^2$$

If we know that the cell of our photometer has an area of say, 2 cm² (i.e., with the circuit in Chap. 13, you have masked the size of the solar cell down to an area of 2 cm² using black tape) and our calibrated output of the photometer is 80 μW, then we know that the power for a single square centimeter will be 40 μW/cm².

Say the manufacturer of our holographic material says that it requires an exposure energy of 160 μJ/cm² . We can look at Eq. 3, which shows that we need to give it a 4-s exposure time.

$$\frac{160 \ \mu\text{J/cm}^2}{40 \ \mu\text{W/cm}^2} = 4 \ \text{s}$$

However, this is only an estimate; and it will probably be on the low side. Remember that we might have objects in the way reducing the amount of light that reaches it further. However, it does help us find the right ball park for the exposure time, and trial and error will ultimately help you make the best holograms.

Project 24: Changing the Color of Your Holograms

You Will Need

- Triethanolamine
- Distilled water

Tools

- Tray
- Regular hologram-making setup

- Hair dryer
- Squeegee

It is possible to make holograms of different colors but using the same laser! This clever trick is accomplished using a chemical called triethanolamine, or TEA. It produces a type of hologram known as a "pseudocolor," or false color hologram. It works by swelling the emulsion, causing it to expand slightly. By changing the amount of TEA in the solution, we can change the amount that the emulsion swells. We then expose the hologram as normal (however, bear in mind that the emulsion has expanded by this point) and then develop the hologram as normal. Bear in mind that the TEA also has the effect of sensitizing the holographic emulsion; this means that you should reduce the exposure time accordingly—usually by halving it.

If you mix a solution with around 10% TEA in distilled water, you will find your holograms take on a yellowish hue. Increase that amount to 15% and they will start to turn green, and by the time the concentration reaches 20%, you will find that your holograms appear blue!

You need to immerse your hologram in the TEA solution. Then, when you remove it, squeegee the

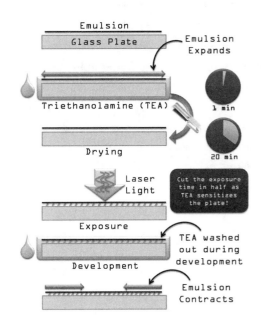

Figure 11-4 *Triethanolamine swells the plate to allow holograms of different colors to be made using the same laser.*

film dry, and give it a wipe with a damp cloth. Ensure that the surface of the hologram is free from spots of moisture or lines because these would appear in the final hologram as lines or spots of a different color. Expose and process your hologram as normal (Fig. 11-4).

Project 25: Chemical Blackening of Reflection Holograms

You Will Need

- Methanol (650 g)
- Water (350 g)
- Sodium borohydride (0.5 g)

Tools

- Plastic tray for immersion of hologram

Up until now, where we have produced reflection holograms, we have blackened the back of them simply by spraying them with black paint. This can present a number of problems because some black paints may react with the holographic emulsion, especially if they are cellulose based (if you are going to use black paint, ensure you pick one with a toluene base). Jeff Blyth has developed a method to chemically blacken the reverse sides of holograms, using sodium borohydride. This formula converts the silver halide into silver through the reducing action of sodium borohydride. Simply mix the water, alcohol, and sodium borohydride together, and immerse the hologram, which must be clean and free from detritus and grease, and completely dry in the solution. The back of the hologram will turn opaque, and at this point you should remove the hologram and wash it vigorously for several minutes to stop the action of the sodium borohydride.

Chapter 12

Computer-Generated Holography

Before we start, I would like to thank Alan Stein for creating and sharing this fantastic open-source tool, which we will be using in this chapter, and for his advice and assistance in the preparation of this chapter.

In the past, optical holography, as we have seen with the experiments so far in this book, was the only way it was possible to produce a hologram. The laborious process of constructing physical models, aligning optics, and ensuring that a perfect exposure of the model was created on a plate or piece of film was the only method available to the holographer.

However, we now live in the digital age. As the power of computers increases rapidly, their ability to create three-dimensional (3D) computer graphics has increased to the point where in the past decade, whole movies can be generated using powerful computers to render the visual output.

The computer has also touched the sphere of holography: 3D computer graphics can be used in place of real physical models, and with some clever computation, it is now possible to produce 3D holograms from master images that exist not in reality but in the graphical memory of a computer.

Project 26: Make Your Own Digital Hologram

You Will Need

- Laser printable overhead transparencies

Tools

- A PC
- A good quality black-and-white laser printer
- Download of the Computer-Generated Hologram Construction Kit (see later)

WWW.

The Computer-Generated Hologram Construction kit is a piece of software that is available for free download from http://corticalcafe.com/prog_CGHmaker.htm.

One of the most failsafe ways that you can execute the computer-generated hologram maker is to open up a terminal window in either Linux or Windows, navigate to the directory into which you have unzipped the files required for the computer-generated holography maker, and type in the following command.

```
java -Xms25m -Xmx250m -jar CGHMaker.jar|
```

A

Figure 12-1   *The letter A, our simple subject.*

This computer-generated hologram was brought to you by the letter A. At least that is what we are going to use as our subject! Because there are limitations to this method of producing holography, let's stick to a nice simple subject. We can see the file, which can be downloaded from the CGH or Evil Genius Web site, in Fig. 12-1.

Software Interface

The software interface for the open source hologram maker is relatively simple to drive. We can see a screenshot of the software in Fig. 12-2.

Xres refers to the number of pixels on the plate in the x-axis.

Yres refers to the number of pixels on the plate in the y-axis.

You will then find a series of three tick boxes. The diffraction grating is an option, which if selected, will not produce a hologram but a diffraction grating. You can learn a little more about these in the chapter on science fair projects, and it's a handy method to bear in mind; but for the time being, we'll leave this box unchecked.

The "Randomize Phase" option can help reduce speckling in the final reconstruction. Try it checked and unchecked, and see what difference it makes.

"Center Object" should be self-explanatory, placing the image or data object you upload in the center of the plate.

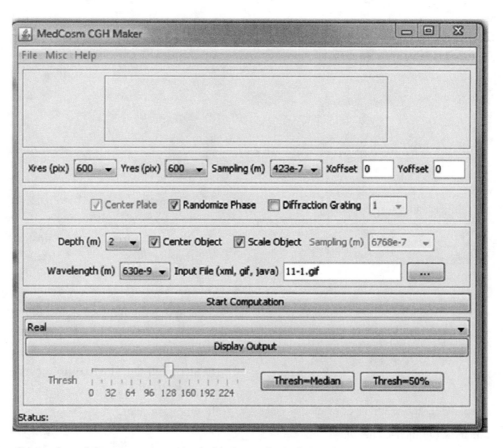

Figure 12-2   *Screenshot of the computer-generated holography software.*

Now we come to another box with some interesting options. The "Depth" control can be used when we are uploading *.gif images only. It controls how far away the object is placed in virtual space with respect to the holographic plate. Experiment by making several different depth computer-generated holograms and compare results.

For the Sampling Rates drop-down menu, consult your printer user's manual to determine the highest resolution your printer can produce, and attempt to match it using these settings.

167e-9	1/4 wavelength @ 670 nm (red)
670e-9	1 wavelength @ 670 nm
846e-7	300 DPI sampling (an older laser printer)
423e-7	600 DPI sampling (a current laser printer)
1692e-7	150 DPI
3384e-7	75 DPI (approximately screen resolution)
6768e-7	37.5 DPI (really bad screen resolution)
1.3536e-3	18.75 DPI

The "Wavelength" drop down is to specify the wavelength of your laser; 630e-9 is suitable for use with red helium-neon and diode lasers.

The input file is exactly what is says! Click the button with the dots to browse your PC locate the file, open the file, and load it.

Inputting Image Data into the Program

We can input data into the program in a number of ways:

- You can define a series of points using XML. There are source files that give examples on the CGH Web site.

- The simplest method, which is the one that we are going to use, is to upload a small *.gif image.

Figure 12-3 *The computer-generated hologram from source file "A" (shown in Figure 12-1). N.B. This image is enlarged and not to final scale.*

- If you are a Java junkie, you can create an object using Java. Again, if you want to explore this option, there are some files available on the Web site.

Once you have set the program up and running you can generate an image which will look something akin to Fig. 12-3. This image has been reproduced larger than you will print it in the final image, so that you can see the fine detail. The detail is computed to replicate the wavefronts that would be formed from a virtual "reference and object beam" hitting our letter "A" shown in Fig. 12-1. If we were producing this hologram optically, these would interfere constructively and destructively, but instead in our computer-generated hologram, this interference is computed, and what is shown on the images is a "representation" of what would be produced on a holographic plate. All of this is done mathematically by the program. In order to view the computer-generated hologram, you will need to print them image "to scale" using a suitable graphics program. Even when printing at the highest resolution possible on a modern laser printer, we cannot come close to the high "resolution" that is attainable through using finely grained holographic film.

Printing Your Hologram

You will need to find a good-quality laser printer. In this application, unfortunately inkjet just won't cut the mustard because we need a dense black print on transparent film. You can get plastic suitable for printing with a laser printer from most office suppliers. You want to ask for "overhead transparency film," and it will look something like Fig. 12-4.

You will get the best results if you can use a lossless image editing program, such as Photoshop, to ensure that the image you are printing is scaled to the right dimensions. You want to produce a 1:1 exact scale image of your computer-generated hologram using your printer's highest resolution settings.

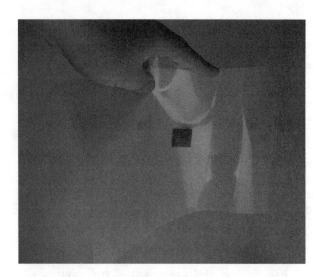

Figure 12-5   *The printed computer-generated hologram.*

Your laser printer will only produce somewhere between one-fifth and one-tenth of the resolution of the holographic film and plates that we use for the other experiments, so quality will be quite dubious (especially when you consider that the resolution is squared, so the lack of data is greatly amplified!) (Fig. 12-5).

Furthermore, the reference beam used in the software is "inline." We've read about some of Gabor's early problems and how off-axis holography improves upon an inline arrangement; however, "off-axis" holography uses more data.

Reconstructing Your Hologram

To reconstruct your "digital hologram," it is as simple as shining a laser pointer through the printed fringes. It is probably best if you can support your laser securely and your hologram securely (you could even fabricate a sandbox mount for these) in order for you to make adjustments and hold your hologram in the right position. A suggested viewing arrangement is shown in Fig. 12-6.

Figure 12-4   *Overhead transparency film.*

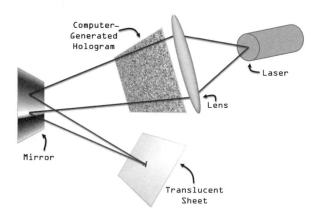

Figure 12-6 *Suggesting viewing setup.*

One way of making all this fit within a reasonable space is to use a mirror to "bounce" the hologram off. A good way of doing this is to take a lens that will focus a diffuse laser beam to a point; you will need a source of laser light (a diode laser with its collimating lens removed will do), a mirror, and some semitransparent screen.

Shine the laser through the lens, position the mirror to bounce the beam back at an angle, and set up the translucent piece of screen at the point where the laser beam converges to a point. Then introduce the hologram just after the lens, and tweak the position of the lens, so the hologram forms a clean image on the screen (Fig. 12-6). With a bit of luck, you should see the letter A.

Do not expect miracles! The computing power to produce these rather low-quality 2D images is quite phenomenal; the images won't "jump out of the page at you." They are a simple 2D reconstruction. The resolution of a laser print on an overhead transparency is such that making the high-quality holograms we make with holographic film just isn't going to be possible.

However, it proves the point that you can generate a computer-generated hologram using relatively simple and affordable equipment.

Chapter 13

Useful Electronic Circuits for Holographers

In this chapter, we present a handful of different electronic projects that will doubtless prove useful to the budding holographer. All of these circuits are relatively simple, requiring only stripboard and "pin through hole" components for assembly and modest electronic skills. To help you make a professional-looking job; I've provided legends that you can stick to a plastic project box and use as a guide for drilling holes for switches, and so on. I suggest you print these legends and then either laminate them or cover them with adhesive clear film. This will help create a project that is durable and survives the inevitable splashes it will receive in the holographer's workshop.

Project 27: Darkroom Timer

You Will Need

- Project case
- 1K resistor
- 10K resistor
- 3 × 12K resistors
- 22K resistor
- 10K preset resistor
- 0.047µF capacitor
- 0.1µF capacitor
- 100µF electrolytic capacitor
- 2N2646 PN unijunction transistor
- 2N2222 NPN BJT transistor
- Small audio output transformer
- Small audio speaker
- PP3 9V battery clip
- PP3 9V battery
- Stripboard

Tools

- Soldering iron
- Drill and bits
- Side cutters

It's often hard to keep time in the darkroom—it's hard to see a clock face by the dim glow of the safelight—and anything too bright runs the risk of fogging your film and providing extraneous light. Help is at hand in the form of this simple audible darkroom timer, as shown in Fig. 13-1.

This darkroom timer will help you time the exposures of your holograms. It provides a tone each second, so there is no light involved that

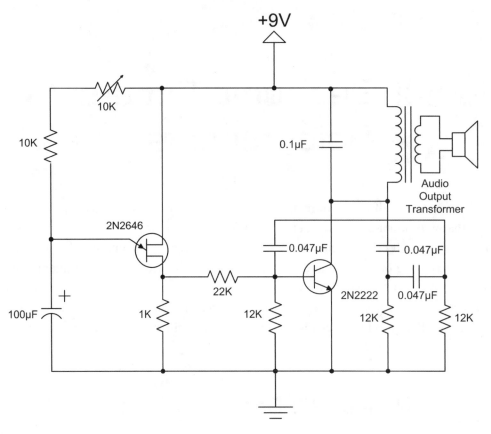

Figure 13-1   *Darkroom timer circuit.*

could potentially ruin an exposure. Count the beeps to work out how much time has elapsed.

If you want to change the value of the 22K resistor, you can. Changing its value will alter the volume of the tone.

The transformer is a small audio transformer, such as you would find in the output stages of a transistor radio. They are cheap and readily available.

The preset resistor adjusts the charging capacitor, which forms part of a relaxation oscillator that determines the timing interval. Adjusting the preset will help you calibrate the timer to keep accurate time.

The legend in Fig. 13-2 will help you make a neat-looking project. It just requires a single hole to be drilled for the on/off switch. This isn't shown in the circuit diagram but is simply connected across the power supply rails! You will see a hole

Figure 13-2   *Darkroom timer legend.*

pattern you can follow where the speaker is to be positioned. Drill each of these holes in turn, and when you are finished, glue the speaker behind.

The finished darkroom timer is a great project and a credit to any holographer's workshop (see Fig. 13-3).

Figure 13-3 *Constructed darkroom timer.*

Project 28: Electronic Shutter

You Will Need

- Moving coil meter
- Length of wire
- Battery

Tools

- Hot melt glue gun
- Scalpel/X-Acto knife or fine-toothed saw
- Small screwdriver set

In this book; the approach we have tended to use with making exposures is to hold a black card in front of the laser, and then when it comes to making an exposure, lifting the card out of the sand, waiting for the relaxation time to elapse, and then raising the card to allow the laser light to reach the optics and then the emulsion. This is all great, but surely there is a more high-tech solution that helps reduce vibration.

If you've ever scratched the surface of photography, you may be familiar with the "cable release." The idea is a simple one. Isolate the camera body from movement, and allow the photographer to control the camera remotely. This is our high-tech equivalent for the holographer's workshop. By ensuring that the holographer doesn't even have to touch the optics table to make an exposure, we deal with one of the prime causes of vibration.

We're going to use a simple moving coil meter to achieve this feat. The concept is a simple one: Apply power to the leads of a meter and the needle of the meter moves. If we attach a small shutter card to this needle, we have an electronic shutter that will move in and out of the path of the laser beam. The meter has its own return spring, so all we need to do is momentarily apply power for the duration of the exposure!

Start with a cheap moving coil meter; such as the one shown in Fig. 13-4. You want to aim for the kind that is usually mounted in the panel of an

Figure 13-4 *Analog moving coil meter.*

Figure 13-6 *Meter with extra plastic cut off.*

electronic project. Don't sacrifice a whole working analog volt/ohm meter (VOM) just for this purpose—that would be a waste.

Then, as we can see in Fig. 13-5, we need to remove the plastic cover from the front of the meter. In this design, the front was held on with a couple of small pieces of adhesive tape. Removing the cover was as simple as running a scalpel blade around the cover, and it duly popped off.

Next you will need to remove the plastic part of the meter that supports the scale (i.e., the meters measurements that you would read in normal operation). This can be done with either a sharp knife or fine saw (Fig. 13-6). You need to be really

careful when you are performing this step that you do not damage the meter needle or the fine armature that moves the needle. You should be left with the armature base and a fine needle that is left unsupported. Don't worry—we're going to attach our shutter to this.

It's up to you how you mount your meter. This will depend largely on the type of holography setup you are using. You need to be careful with magnetic mounts because these could compromise the working of the moving coil that is needed to actuate the shutter. In this instance, I've simply glued the moving coil onto a foam board base as you can see in Fig. 13-7. This is quite useful because it can then be used with optical mounts

Figure 13-5 *Meter front case removed.*

Figure 13-7 *Meter hot melt glued to piece of foam board.*

with "clips" to clip the meter into place. You will see also that a small square of foam board has been cut as our shutter card. The beauty of foam board is that you can readily stick sharp objects into it. So take the delicate meter needle and stick the square of foam board on the end. In any case, if you find the fit is a little loose, a dab of hot melt glue will improve the situation.

Operation of the shutter is relatively easy. It's just a matter of applying a voltage to the wires that lead to the meter. If you are going to use the shutter with the automatic timer circuit presented next, then make sure that when you remove the voltage, the strength of the spring is sufficient to move the little shutter card back to the resting position. If you find this action is a little sluggish, try trimming the shutter card to reduce its weight.

You can also add a momentary double-pole double-throw (DPDT) toggle to reverse the direction of the voltage to the meter. If you find that the shutter card tends to stick in the "activated" position, a brief pulse of reverse-polarity voltage should send the shutter card flying the other way! You can see the two positions of the shutter card in Figs. 13-8 and 13-9.

By using this shutter with a small battery and switch, you can make holographic exposures free of worrying that your hand is going to cause vibrations in the sandbox.

Figure 13-8 *Shutter in resting position.*

Figure 13-9 *Shutter in activated position.*

Project 29: Automatic Electronic Shutter

You Will Need

- 555 timer integrated circuit (IC) (Note: A socket to mount it in is handy and saves ruining an IC if you solder poorly.)
- 1M variable resistor
- 1000μF capacitor
- 100F capacitor
- 10F capacitor
- 1μF capacitor
- 100K fixed resistor
- Rotary switch
- Pushbutton switch

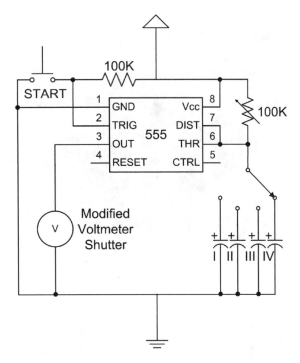

Figure 13-10 *Automatic electronic shutter timer circuit.*

- Small moving coil voltmeter (as in previous project)
- Speaker terminals
- Project case

Now we can take the previous project a step further by automating it. With the help of a simple timer circuit shown in Fig. 13-10, we have the ability to time our exposures accurately and in a way that is repeatable. This is important because if we want to make consistently good holograms, we want to ensure that we can record what works and what doesn't.

Hint

The circuit should appear to be simplicity itself. We're using a 555 timer to actuate the shutter from the previous project. As with all projects, although not shown on the circuit diagram, it is worth including an on/off switch if you are going to put it into a project enclosure. It makes a better job than continually removing the batteries.

The circuit is configured as a monostable timer; when the 555 IC is triggered, it will raise its output high for the time period specified, moving the electronic shutter, and then at the end of that period, the output of the 555 will turn low, returning the shutter arm to its resting position. There is a continuous adjustment that can be made by turning the 1 M variable resistor, and the range of times can be selected using the rotary switch to select the capacitor. Note: To keep to round numbers and easily obtainable components, at its fullest extent, the timer will in fact time up to 10% more than indicated on the range. However, by calibrating the timer against a known time source, marks can be placed on the continuous scale to indicate what position the variable resistor must be in for 1 to 10 and all the numbers in between.

The following table shows the values of capacitor that should occupy positions I to IV in the circuit diagram.

Position Number	Capacitor Value	Time Range (s)	Maximum Time (s)
I	1000μF	0–1000	1100
II	100μF	0–100	110
III	10μF	0–10	11
IV	1μF	0–1	1.09

You want to source the sort of voltmeter that is used for small voltages and is "panel mount." See the previous project for some notes. A signal strength meter from a radio/tape deck is a good choice and would be easy to salvage from old junk. It is essential that it is an "analog" meter. We're not too worried about how accurate it is or what voltage range (as long as it is similar to the output of the 555). If your meter is calibrated to measure voltages that are much smaller than the output of the 555, you can add a resistor in series with the meter to step down the voltage. A variable resistor would allow you to start from a high resistance and then adjust for maximum deflection when the circuit is activated (but no more—in case you break the meter).

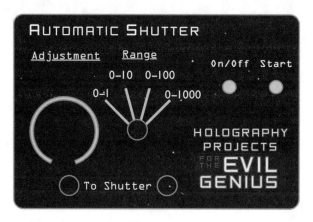

Figure 13-11 *Automatic shutter legend.*

Figure 13-12 *Automatic shutter constructed.*

Looking at Fig. 13-11, we can see the two circles at the bottom of the picture that indicate where the terminals should go for the shutter itself. When I assembled this, I used speaker terminals. They have the handy property that they can be used with both banana plugs if you want to terminate the tails of the electronic shutter in a neat and professional manner. Or they can be used with bare wires that can be clamped under the screw terminals.

The two switch holes for the "on/off" and "start" switches should appear self-explanatory.

There are two more holes that must be drilled. The range hole must be drilled to accommodate a rotary switch. We're going to give some more details later on about how rotary switches work and what you need to know. Cut the shaft of the switch down and affix a nice plastic knob, and you'll find the job looks that much smarter!

Adjacent to the rotary switch is the "adjustment"; because this works across several different scales of operation, it is left intentionally blank. In operation you will come to judge what positions give you timings of which length, and you can use the audible timer to help calibrate this scale if you like. The adjustment needs to be drilled to accommodate a variable resistor. And again, if you go to the trouble of trimming down the shaft and screwing a proper knob on the end, the result will be a much more professional-looking project.

You can see the final assembled circuit in its box in Fig. 13-12—a useful piece of darkroom equipment!

Project 30: Simple Photometer

You Will Need

- 741 integrated circuit (Grab yourself an 8-pin DIL socket if you aren't confident with your soldering to safeguard the IC).

- DPST switch
- Rotary switch
- 2x PP3 battery clips
- 2x PP3 9V batteries
- Voltmeter (measuring around 10V preferable)
- 15K resistor

- 100K resistor
- 4x multi-turn preset variable resistors adjusted to values below
- Solar cell (see following discussion)

- Soldering iron
- Drill
- Side cutters

In this circuit, we use an OpAmp to amplify the signal from a solar cell and display its output on a volt meter. We also have the function that we can select between four different values of resistor, which will change the gain of the OpAmp to allow us to read the output from the sensor on different scales. If you can borrow a calibrated photometer, you can calibrate yours against it by marking the scale of the meter. A small cheap meter is useful for making basic comparisons, a larger meter with a more detailed scale has the potential to be an accurate and useful instrument around the workshop.

Operational amplifiers require a dual rail power supply. That is, they require one supply line above ground potential and another below. To accomplish this we use a pair of 9V batteries. Never try and operate this supply with just a single battery in, and ensure that the switch is set to off when changing batteries. The on/off switch will cut both supply rails because it is a DPST switch. Use the table to calibrate the variable resistors that are used in this circuit, and follow the circuit diagram in Fig. 13-13.

Switch Position	Resistance Value	Meter Sensitivity
I	1.8K	0.1μW
II	3.6K	0.5μW
III	18K	1μW
IV	180K	10μW

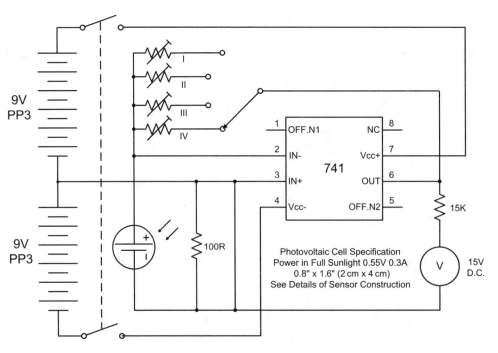

Figure 13-13 *Photometer circuit.*

We need a "photocell" that will detect the amount of light incident on the hologram; in this instance, we want a small solar cell. This solar cell needs to be mounted in an enclosure. Although we could go to the trouble of sourcing all of these components individually, we can get them all cheap as an off-the-shelf item if we're prepared to be ingenious and hack something off the shelf. You may have seen the cheap solar torches. They contain a white light–emitting diode or two, a small battery, and a small solar cell. The idea is you leave them on your window sill until you come to use them. Then pick them up and they are charged with the sun's energy. If you can bear to cannibalize one of these, you have an off-the shelf solar cell mounted in a neat enclosure. The sort of thing I am talking about is shown in Fig. 13-14.

To hack the torch, simply use a small delicate screwdriver to take it apart. The switch, battery, and driver circuitry should all be readily apparent. A few deft snips with a pair of side cutters and

Figure 13-15   *The completed photosensor.*

possibly a screw or two, and your experimenter's parts bin acquires some new components. You are left with a solar cell and a neat plastic box! If you want to make a neat job of it, drill a hole in the back of the box and fit a small strain relief where the cable exits. You will need to mask off half of this cell with some black opaque material. A piece of plastic or black foam board will suffice. Fig. 13-15 shows the completed photosensor.

The device will produce a reading on a voltmeter in the x1 mW range (1 mW equals 1 volt). You can confirm this by taking your laser, aiming it at the exposed part of the photocell, and looking at the reading. Although the meter can be used to take rough absolute readings, arguably it is more suited to taking relative readings of beam intensities, for example when evaluating the strength of reference and object beams in a split-beam setup.

As we have presented the project here; it is a simple meter that due to its small size is only really useful for making relative comparisons between different beams. If you were to use a bigger meter, with a clearer scale, you could calibrate the device to allow you to make absolute measurements of light intensity against a common reference (say, measuring your laser beam spread at a given unit distance). If you want to go down

Figure 13-14   *Solar torch for modification.*

Figure 13-16   *Multi-turn preset resistor.*

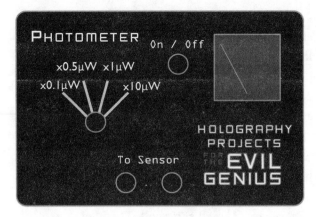

Figure 13-17   *Photometer legend.*

this road, then try and ensure that you build your circuit using components with values that are as accurate as possible. When it comes to setting the values for your four resistors, a little extra investment will get you a "multi-turn" preset resistor. You can see one of these in Fig. 13-16. Rather than crude presets, which only allow you to get somewhere near to the right value, a multi-turn preset is variable over its range, but rather than covering the range in a single turn, by its nature a multi-turn preset resistor requires multiple turns, which allows you to get much closer to the value of resistance that you want to achieve. My advice is that before you solder them to your printed circuit board, set the values that you need using a fine jeweler's screwdriver, mark the casing of the multi-turn resistor using a permanent marker, to indicate what value you have set it to. Then apply a little thread lock or hot melt glue to the screw, to ensure that the resistor remains set at that value.

As with all of the projects in this chapter, a handy legend reproduced in Fig 13-17 is provided for you to print and use on your project case. As with the previous project, you might want to

consider using speaker terminals as the contacts for your photosensor because you can use them with both banana plugs and bare wires. There is a space for an on/off switch and a rotary switch to select the range of the meter. If you rewind back to Fig. 7-15, you can see a similar photometer produced by the Open University in small quantities to accompany a distance learning course that included holography as one of the activities. The circuit appears broadly similar to the simple circuit presented here, albeit with some additional options. The larger scale meter is calibrated to switch to select the range of the meter, you can see the finished constructed meter in Fig 13-18.

Figure 13-18   *The final constructed photometer.*

Hint

To calibrate the meter, adjust the area of the solar cell that is exposed by changing the position of the opaque mask. It is helpful if you can obtain a calibrated photometer to benchmark this one against because then you can make more accurate readings. For low light levels, it is permissible to remove the mask from the photocell and use the whole area to make relative comparisons.

The scale on the meter becomes arbitrary because we are measuring light over a few different scales. However, I highly recommend buying a package of "permanent" pens that are used for marking on overhead transparencies. Use these to place "dots" on the plastic fascia of the meter that correlate to important measurements. Use a few different colors (but remember which dots are which color when you are under safelight!). You can make some meaningful measurements that will allow you to make judgments as to the relative amount of light incident on the film plane.

Hint

Notes on Rotary Switches

Many rotary switches follow a standard modular design where a standard switch body is available in a number of different variations. Rotary switches typically have between 1 and 4 contacts in the middle, on the reverse of the switch, and 12 contacts around the outside. When you buy your switches, they are described by the number of "poles" they have—which is to say, the number of common contacts that can be switched between a number of different "ways." In one common design of switch; the number of "ways" is 12 divided by the number of "poles."

When you remove the locking nut that is used to secure the switch into an enclosure, you will see a metal washer with a bent tab. Remove this, and underneath you will see a series of holes, with numbers molded into the plastic body of the switch. This metal locking tab can be used to limit the number of "ways" on the switch that can be used. In our case, it's irrelevant how many "poles" our rotary switch has; one pole will do just fine. But for both of our projects we want to use the metal locking tab so it has four "ways." Or to put it differently, so there are only four possible positions that the switch can be set in. Figure 13-19 shows the locking tab removed.

Furthermore, in Fig. 13-20, you can see the arrangement of the terminals to the reverse of the switch.

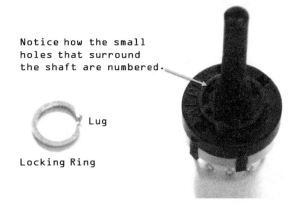

Figure 13-19 *The locking tab removed from the rotary switch.*

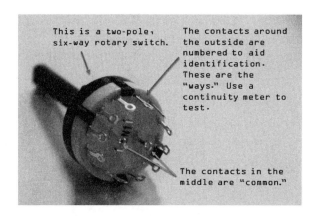

Figure 13-20 *Rotary switch wiring details.*

You Will Need

- Tricolor LED
- 100 ohm resistor
- 2x AA battery holder
- 2x AA batteries
- Center-off changeover switch
- Soldering iron

You Will Need

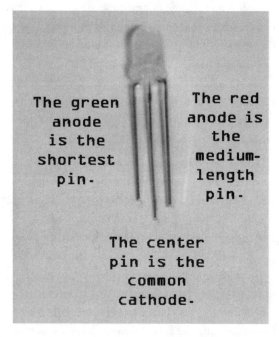

The green anode is the shortest pin.

The red anode is the medium-length pin.

The center pin is the common cathode.

Figure 13-21 *Connection details for a tricolor LED.*

This circuit is too simple to warrant a circuit diagram; however, if you are unfamiliar with them; there is a brief explanation of how a tricolor LED should be connected in Fig. 13-21. A tricolor LED is a red and a green LED integrated into the same package. They are available in common anode and common cathode variants, but they both perform the same function. Sometimes when we are working with "red"-sensitive emulsions, we require a dim green light to work by. At other times, when we are working with "green"-sensitive emulsions, or for example if we were making holograms using dichromated gelatin, a red safelight is called for. In this book, we're not going to cover using "green-sensitive" emulsions, but it

is useful to have that extra functionality there as your hobby progresses. Ensure that you include a resistor on the common lead, and use the changeover switch to switch between the two colors of safelight. Remember, it is always worth testing your safelight against a scrap of holographic emulsion to check that it is not "fogging" your film. If you find that the safelight is helping to expose your film, move it further away, try adding a diffuser, and, as a last resort, check that the wavelength of safelight you are using is appropriate for the emulsion of film you are trying to use.

Chapter 14

Science Fair Projects
for Holographers

If you've persevered this far into the book, the chances are you have already got to grips with the fascinating science of holography; and I'll bet that you're hooked and want to go further with it!

Holography is such a rich subject area for any budding amateur scientist; it brings together diverse strands of physics, chemistry (and in the mushroom experiment in this chapter, even biology), and it is a good platform to allow you to demonstrate some of the science knowledge you

have acquired. Although making small holograms is fascinating and allows you to share your expertise with your friends in an artistic form, there is nothing like discovering something new or taking the knowledge you have applied and finding new contexts for it.

The first project, if you've followed the signposts in the book, is useful to build before you have even set about making any holograms. It will help you to identify sources of vibration that can ruin your holograms!

Project 32: Michelson's Laser Interferometer

You Will Need

- Laser
- Beam splitter
- Two mirrors
- Piece of white card
- Lens

In this project, we are going to build a Michelson Interferometer. You can see the basic setup for the interferometer in Fig. 14-1.

The Michelson interferometer records changes in the position of the optics used in the project, and

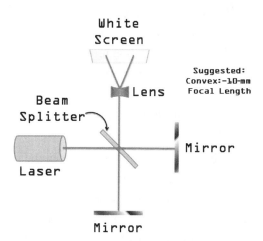

Figure 14-1 *Michelson interferometer.*

so by looking at the changes in the position of the optics in our sandbox, we can identify what vibrations are present that will cause them to move.

This instrument is so sensitive, it can even measure the vibrations caused by the quietest of whispers!

Building the Interferometer

Sand is fantastic because you can draw temporary lines in it to help guide your layout. Draw a cross in the sand, with the two lines at 90° to each other, splitting the sandbox into four quadrants. Along the "long edge" of your sandbox, position your laser so it points along the longest line. Aim the beam of the laser straight across the sandbox, and at the other end, get a mirror, and bounce that beam back to the laser. Make sure they are at opposite ends of the sandbox. You want to reflect the dot of the laser beam back so it hits the point immediately above the aperture where the beam emerges.

Get your beam splitter, and at the point where the two lines cross, place it at a 45° angle—your laser beam will pass through the splitter. You might need to mask off the beam splitter to ensure there are no secondary reflections. Now you will have two dots that should hit the sides of your sandbox round about where the short line touches the sides. Looking at the beam splitter, you will see there is one side where the beam coming from the laser shines first off the surface of the splitter. Here you want to place a mirror to bounce the beam back to the beam splitter. On the opposite side of the sandbox, you will see two dots. (You might even see a few more caused by secondary reflections, but ignore these.)

Set up a diverging lens, and place it in the path of the two beams. By tweaking the last mirror that you set up on the short arm of the cross, you should be able to get the two beams to align, producing (if you stick a white card in the sand some distance from the diverging lens) an interference pattern.

Your setup should look something like the aerial view in Fig. 14-2 and the side view in Fig. 14-3.

Figure 14-2 *Michelson interferometer aerial view.*

Figure 14-3 *Michelson interferometer side view.*

The interference pattern will look something like Fig. 14-4. You will see fringes of light and dark, and unless you are keeping rather still, the chances are they will be moving about! Keep everything still and you should see them stabilize. Experiment with creating different types of vibration. Talk to one of the optical elements: The sound of your voice will make it vibrate.

Sometimes, if you're in an area where there is heating or airflow, you will see a slow movement back and forth of the fringes. Attempt to isolate the draft or air currents!

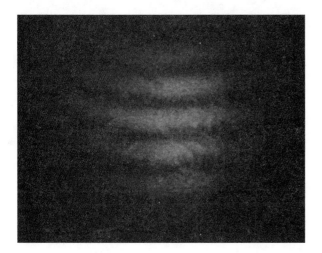

Figure 14-4 *The interference pattern generated by the interferometer.*

Vibration Testing Your Sandbox

You need to ensure you are working with a rigid sandbox before you even set about making any holograms. The slightest vibration is the difference between a brilliant sparkling three-dimensional image and a dull blurry plate with nothing recognizable recorded. When you think that a helium-neon laser produces light with a wavelength of 0.0000317 mm and that a disturbance even this small could ruin your hologram, then you'll realize how crucial vibration isolation is. Any construction will have what is known as a *relaxation time*, the time taken for it to settle after it has been agitated. To work out how long your sandbox takes to relax; first wait until the lines on the screen are stationary. Then gently push a corner of your sandbox and release. The fringes on the screen will jump about a little, but using a timer, you should be able to determine how long it takes for the fringes to come to a standstill. Once you have taken this measurement, you are aware of the relaxation time. You should allow this time before making an exposure every time you adjust, move, or knock the sandbox or optics bench. You will find that some different vibrations will result in different relaxation times depending on their intensity.

Project 33: Laser Interferometry

You Will Need

- Setup for standard holography exposure

Tools

- Ruler
- Other plastic objects
- Small weights (coins are ideal)

Engineers often need to test objects to analyze potential failure mechanisms and weaknesses. A good example can be found in car tires; vehicle safety is incredibly important, and it is essential that tires are able to perform under a wide range of conditions to ensure vehicle safety because a tire blowout could potentially result in loss of life.

Any engineering component that has the capability to deform or move under pressure is ripe for analysis by laser interferometry. The basic method is this. Two holographic exposures are made with the object, first in its undeformed state and then in a slightly deformed state. You want to pick an object and weight such that the movement is barely imperceptible to the human eye. However, when the hologram is developed, it will reveal all!

For a simple setup, let's just use a standard stationary ruler, the sort you're likely to have in your pencil case. Once you've done this simple experiment, you can move up to small plastic models and the like.

We are going to be carrying out a process called *double-exposure interferometry*. We make one exposure with the ruler "unstressed" and then apply a tiny weight, to cause imperceptible bending forces. Set your equipment up as shown in Fig. 14-5. However, the hologram will reveal all. Make one exposure but make it short—you only want a "half exposure." Then apply the weight gently without disturbing anything else. Make a second exposure, again for half the time you would normally take.

What you will see is interference fringes on the ruler, indicating where it moved. Nothing else should have an interference fringe. If it does, you were careless and knocked some other equipment during the exposure. Each fringe will represent a displacement half a wavelength, with the first one showing a displacement of a quarter

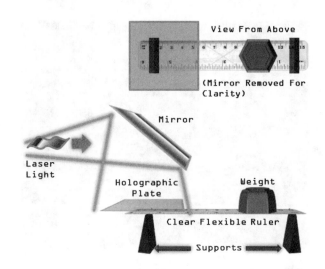

Figure 14-5 *A laser interferometry rig analyzing a ruler.*

of a wavelength. As we know, a helium-neon laser has a wavelength of 633 nm, we can work out the tiny amount by which the ruler deformed!

Taking it Further

One customer of Integraf used this method to nondestructively test an F-16 Jet. Admittedly, it was a plastic model, but the ambition was there.

Project 34: Watching Mushrooms Grow

You Will Need

- Mushroom growing kit (from your local garden center)
- Setup for standard reflection hologram

This sort of holography project should appeal to a "fun-guy." Appalling jokes aside, as we saw

in the last experiment, we can do interferometry with things that deform or move slightly between two concurrent exposures. Now we don't want things to move *too* much. As we have seen from holography fault finding, this can result in piece of hologram missing from the exposure. So we need to find things that deform or move very slightly.

Mushrooms grow very quickly, so quickly in fact that with some kinds of mushrooms and some clever holographic trickery, we can even see how they grow by capturing two holographic exposures with a time separation in between. This then opens up a whole vein of exciting science, where we can analyze the factors that affect mushroom growth—heat, light, different nutrients, different growing media, different varieties of mushrooms—but hey, this is your science project, so you go think of some more variables to experiment with!

This was a prize-winning science fair project of one of Integraf's customers, so it shows that with the right execution, holography projects can support you all the way in school science.

The great thing about this project is that it requires the same simple materials that you've been using to make some of the most basic holography projects in this book. All that you need to add are the mushrooms.

Now mushrooms can prove a tricky subject to catch a good hologram of because they are soft and deform easily. Furthermore, if you are going to experiment with factors such as changing temperature, you need to think about how you can isolate this effect to the mushrooms and growing media. Changes in temperature to the hologram can affect the fringes that form the hologram image and ruin the hologram you are recording.

When you buy a mushroom from the store, you're not buying the whole shebang! What you're actually getting is just the "fruiting" part of a whole network of little filaments and strands that colonize the substrate (the material in which the mushroom grows) in order to extract nutrients and nourishments for the mushroom. Traditional mushroom compost was made from horse manure and straw, but more modern substitutes are available now.

Mushrooms fall into two classes: those that grow in the ground and those that grow in a host piece of decaying wood. The easiest way into mushroom growing is to look at your local garden center where you can find kit containing everything you need—the right sort of compost, spores, and so on, to grow your mushrooms. These kits provide a good base for experimentation. They provide you

with a known "control" that is proven to work. From this starting point, you can deviate and change variables one by one to see how it affects mushroom growth, which can be measured over a short period using holographic interferometry.

With the ground growing varieties of mushrooms, a mushroom growing kit generally comprises a container, some compost designed for the job, and a packet of "grain" that has been treated with the variety of mushroom specified in the kit. It's just a matter of sprinkling the grain over the compost, watering well, and keeping them in a sheltered location.

With the varieties of mushrooms that grow in decaying wood, the spawn are supplied in the form of wooden plugs. You need to find a hardwood log to act as the host. It should have been cut within the preceding month and a half. It needs to be substantial if you are to secure the full mushroom crop. Using a drill, bore some holes in the log of a sufficient size to insert the plugs. These should fit snugly. Mushrooms will appear within the next several months.

Mushrooms do not tolerate extreme variations of temperature. If they are too hot or cold, the growing process will stop.

Once you've successfully got some mushrooms growing, you need to repeat the process that we saw in the previous experiment, making two exposures. Figure 14-6 shows an example of how the

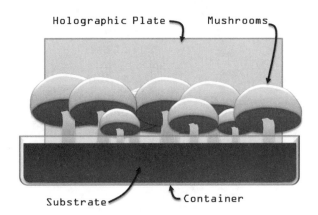

Figure 14-6 *Two exposures made on a holographic plate alongside growing mushrooms.*

holographic plate can be set up to capture the mushrooms growth from the side. However, this time because the mushrooms are doing the movement for us, time them. You could make different holograms, each with a different exposure time, and using the method before, calculate the movement.

Project 35: Experiment with Diffraction Gratings

- Laser
- 3x mirrors
- Beam splitter
- 2x spatial filters
- Holographic plate

Holographic optical elements (HOEs) are becoming an increasing sight in many modern optical devices. The essence of HOEs is that a hologram can be made of an optical device that will then replicate the properties of that optical device.

A diffraction grating is the flat holographic equivalent of a prism. We've all played around with prisms of glass, shining white light through them to find what emerges from the other side is a spectrum of rainbow colors. It's incredibly easy to make a device that will fulfill the same function as a "prism" from a regular holographic plate, using the setup detailed in Fig. 14-7.

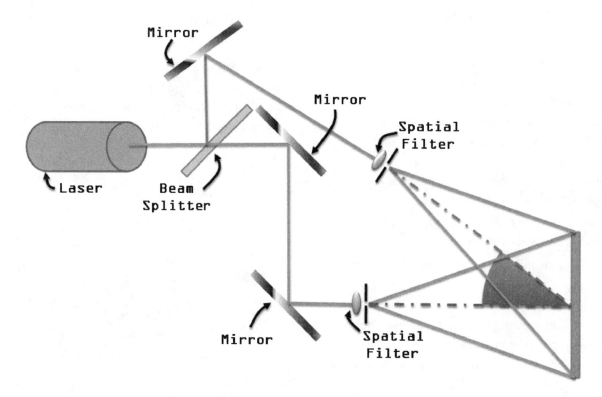

Figure 14-7 *Setup to make a diffraction grating.*

If we make a prism with a red laser—say a helium-neon laser or a diode laser, then when we shine white light through it, the red component of that light will be diffracted at the same angle as the original beam; however, the shorter wavelengths of light (all the other colors) will be diffracted at shallower angles, relative to the frequency of the light. This manifests itself as the prism-like property of spreading the light out in a rainbow, separating it into its component colors.

A diffraction grating is effectively a dual-beam hologram made without an object! Expose the hologram, develop it normally, and you should find that your diffraction grating has prism-like properties that you can then experiment with.

Hint

If the angle between your two beams is less than 45°, you will find that when you use your diffraction grating, you will get higher order diffraction. What this means in principle is that when you shine a laser through your grating, you will get more than one dot to emerge.

www.

If you are interested in taking this further, then check out Integraf's Web site, which has a list of holography-related ideas that you might want to try! http://www.holokits.com/a-hologram_ science_ project.htm.

Chapter 15

Other Non-Holographic Three-Dimensional Projects

In this chapter we will tear down all the different technologies that although don't pass the test as far as technically being classed as "holograms," are often called holograms by the ill-informed. We're going to look at what makes holography different from other three-dimensional (3D) imaging technologies and discuss some of the pros and cons of different methods of making images "jump out from the page."

Partly because of the influence of science fiction, the term *hologram* has been used and abused to come to represent a wide variety of technologies that can produce 3D and some technologies that don't yet even exist except in the minds of science fiction writers and special effects producers.

Simple 3D imaging isn't a new idea. Very quickly after photography was developed, stereographers discovered, that by taking a pair of pictures, a 3D scene could be "replayed" to the viewer with a simple viewing apparatus or a trick of the eyes. Charles Wheatstone realized this as early as 1838; however, his early stereograms were simple drawings not photographs. Shortly after in the 1840s, the science of stereography began to really gain momentum, and in 1851, Queen Victoria viewed and praised the technology of stereography, which she viewed at the Great Exhibition in 1851. Over on the other

side of the Atlantic, Oliver Wendell Holmes developed a simple viewer that could be used to view stereographs, consisting of a pair of lenses and a support for a picture at a fixed distance (a modern-day reproduction in cardboard of a stereo viewer can be seen in Fig. 15-1). And by creating vast libraries of stereo images and mass producing stereograms, the novelty became an important tool for education and a toy for entertainment.

Now let's look at how you can create and view your own 3D images.

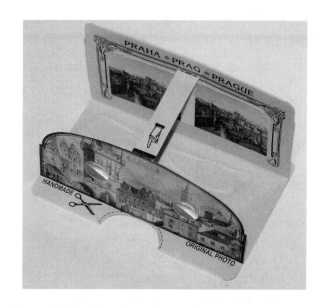

Figure 15-1   *A modern-day cardboard stereo viewer.*

You Will Need

Method 1:

- A stereo camera (e.g., Loreo 3D)
- 35 mm film to suit camera

Method 2:

- A single-lens reflex camera with changeable lenses
- A stereoscopic converter (e.g., Loreo 3D lens in a cap), also sometimes referred to as a "beam splitter"

Method 3:

- A camera with a tripod mount
- A tripod
- A slide bar

Method 4:

- Homemade inexpensive stereo camera (see next project)

Hint

If you peruse the online auction sites using search terms such as *stereography, stereo photography, 3D photography,* and so on, you will doubtless be able to find a plethora of vintage camera equipment. There are some vintage 3D camera available at a reasonable price; however, if you are looking to buy new equipment, undoubtedly the largest current supplier of such equipment is Loreo, which can be found online at www.loreo.com.

You may remember from Chap. 1 that the reason we see in three dimensions is because of the separation difference between our eyes. *Stereoscopy* is the art and science of taking a pair of images that can then be viewed individually by each eye to create a 3D effect. Stereoscopy is distinctly different from holography in that it uses two "focused images," whereas holography uses an "unfocused" image that is made with coherent light. A hologram records all of the 3D information falling on the film plane from the angle it is at relative to the object, whereas a stereo pair of images only records the image from a static viewpoint. With a hologram, you can move left and right and up and down, and the image in the picture should move accordingly—your viewpoint will change. However, a stereo image is static. You can only see the scene from the position of the camera when the stereo pair was created.

There are a number of ways that we can create stereo pairs.

Method 1

The simplest way is to buy a "stereo camera" made for the job. Currently, Loreo produces a basic point-and-click stereo camera (see Fig. 15-2).

Figure 15-2 *Loreo stereo 3D camera.*

The camera is relatively affordable and accepts 35 mm film. The camera uses two lenses and a complicated mirror and prism system to take two views and then capture them side-by-side on a standard frame of 35 mm film. This system has some inherent advantages. As far as film-based cameras go, it is simple because it allows you to take the film to an ordinary processing outlet and have it developed and printed in a standard size. Loreo then sells a viewer (shown in Fig. 15-8 later in the chapter) that will accept standard sized 6 × 4-in pictures for viewing. As a cost-effective, simple, and reliable entry to the world of stereo photography, this system is quite unbeatable. You might also want to browse the online auction sites for some "vintage" stereo photography equipment. There is quite an array of different equipment available that will shoot two pictures on standard film sizes. Some of these cameras will take "full-frame" stereo pictures using two whole frames to shoot each 3D stereo view. This has the disadvantage that you use your film twice as quickly, but it is good for quality work.

Method 2

The next option, which is worthwhile if you already own a single-lens-reflex camera with interchangeable lenses, is to buy a "stereo lens." There was a fad for stereo lens devices, and there are some different secondhand vintage models available from specialist photographic dealers and the Internet. But again, if you are buying new, Loreo has cornered the market with its simple "lens in a cap" system that is available to fit major camera lens mounts. You can see this in Fig. 15-3. There are versions available for both standard film cameras and digital cameras, which as a rule tend to have a sensor that is smaller than the size of a 35 mm frame of film. The "lens in a cap" does a similar thing to the optics in Loreo's point-and-shoot stereo camera, albeit using your camera body.

Figure 15-3   *Loreo stereo 3D "lens in a cap" for SLR camera.*

Method 3

Finally, in the realm of "professional solutions," you can buy a system which affixes to your tripod mount and allows your camera to "slide" along a rail. Figures 15-4 and 15-5 show top and bottom views of the sort of rail that can be used with a camera and tripod to take two subsequent exposures that are horizontally transposed. To use this system you must use a tripod (it's also useful if you buy a small spirit level for your camera to ensure that it is horizontal when you are taking shots; these are available from photographic suppliers, and there is even a model available which affixes to the "hot shoe" socket where your flash would conventionally mount).

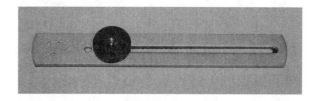

Figure 15-4   *Camera slide bar for tripod mount (top).*

Figure 15-5   *Camera slide bar for tripod mount (bottom).*

Two exposures are taken, moving the camera along the rail between the exposures. This system makes sense if you want to make views of static subjects, such as buildings or landscapes; but isn't suitable for subjects where there is likely to be movement between the two exposures.

Method 4

The next option, while rough and ready, is a dirt-cheap entry into the world of stereo photography and a fun project if you wish to explore this world.

Project 37: Making an Inexpensive Stereo Camera

You Will Need

- Two disposable cameras
- Roll of adhesive tape

This project really is incredibly simple, yet the results are astoundingly satisfying. Take a pair of cheap ordinary disposable cameras, and using some adhesive tape, join them base to base, as you can see in Fig. 15-6, so that their lenses, when the camera is held in portrait orientation,

Figure 15-6 *Two disposable cameras taped together to form a simple stereo camera.*

align horizontally. This may mean that when you join the cameras, they are not perfectly square, but do not worry. The key feature is that the lenses are in line.

You will want to ensure that you select two cameras of the same brand, with the same style of lens containing the same type of film. Try and ensure that they are identical in every respect.

Why do we join them base to base and not side by side? Well, joining the cameras side by side would result in the lenses being positioned dramatically wider than the separation of our eyes, so the result would be an "exaggerated" stereo effect that our eyes would find hard to resolve.

You can only use this as a 3D camera if you hold it so the lenses are in the horizontal axis, taking two side-by-side images. When holding the camera, you will need to use two hands, to ensure that both shutter buttons are pressed simultaneously. This is a little tricky, but you can see an example of how to hold the camera in Fig. 15-7. Ensure you keep your fingers far away from the lenses and flashes when taking pictures.

One of the things you will need to remember with this design of camera is that once you have taken a picture with both lenses, you need to ensure that you advance both films at once. If you forget, you will find with your next

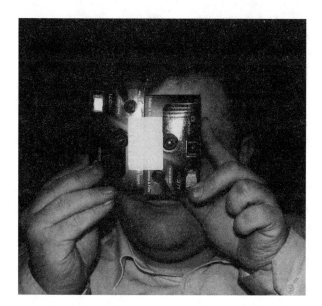

Figure 15-7 *Taking a picture using the "disposable" 3D camera.*

exposure that only one shutter fires, and you've wasted a shot.

When it comes to getting your pictures processed, ask the lab to process both of the films using an identical process. You want to ensure they are treated in the same way so the prints you get from each camera are as similar in every way possible, except for the small detail that they are taken from slightly different viewpoints.

Label your two cameras "left" and "right," and when it comes to getting the images processed, make sure you keep a note of what set of pictures is what. If you can afford to "waste" an exposure, the simplest way to do this is with the left camera. Take a picture of your left hand flat on a desk, and with the right camera take a separate picture of your right hand on a desk. Then when you receive your two packets of pictures back, the first exposure of each film will remind you what set of images you are looking at.

When you receive your two sets of images; you will need to join the views from left and right, together in pairs to form your "stereo views," which can then be viewed using some of the devices and gadgets shown in this chapter.

Hint

Another cheap way into stereo photography is with a kit from the Japanese manufacturer Gakken that sells a stereo pinhole camera. This takes the concept of a pinhole camera that we first met in Chap. 2 and unifies it with the theory of stereo photography, to produce a relatively cheap kit that uses two pinholes to make a pair of stereo images on standard 35 mm film. There are a couple of places online where you can buy the camera.

1. http://www.makershed.com/ProductDetails. asp?ProductCode=MKGK16&Show=TechSpecs

2. http://www.verycoolthings.com/vct/Neo_ getpage. cgi?page=itemtoy&itemID=14838

Typing "Gakken Stereo Camera" into the search box of a site like Flickr.com will yield a plethora of photos that have been generated by users of the Gakken stereo camera, and there is even some discussion at http://www.flickr.com/groups/898882 @N20/discuss/72157612734843409/ that includes links to the instruction manual of the camera. Unhelpfully (unless you're a Japanese speaker) these are written in Japanese; however, the pictures are informative and useful. If you are handy with an automated translation service such as "Google Language Tools" or if you are a Japanese speaker, you may also find the Web site of Gakken useful: http://otonanokagaku.net/magazine/ vol14/index.html.

Viewing Your Stereo Images

In Chap. 2, you will find a method for free-viewing stereo pairs; however, if you have the budget to buy a stereo viewer, you will find that it makes the whole experience much easier. Again, Loreo is the big name in 3D imaging if you are planning to buy new. They sell two different varieties. Their

Figure 15-9 *Loreo Lite viewer.*

Figure 15-8   *Loreo deluxe viewer.*

"deluxe" viewer (see Fig. 15-8) is made from rigid plastic. It is ideal if you are planning to make a hobby of 3D photography. However, if your pockets are not that deep, then a "Loreo Lite" 3D viewer (shown in Fig. 15-9), which contains two lenses and can be folded simply from a card,

represents a cheaper investment. Although cheaper and of simpler construction, the Loreo Lite viewer does have some advantages. Because you have to hold it away from the stereo pair of images yourself, it is useful if you get a pair of standard 6 × 4-in photographs from your disposable 3D camera. These will not readily fit in the deluxe viewer without some "scanning and scaling" to fit. Furthermore, if you anticipate working on your 3D images on a PC, the Loreo Lite viewer is useful.

Project 38: Digital Stereo Photography

You Will Need

Method 1
- Two cheap digital cameras

Method 2
- Fujifilm Digital Camera Fine Pix Real3D

Method 3
- Minoru 3D Webcam

Method 1

Traditional film-based cameras have some serious limitations. It's hard enough to compose a decent picture in two dimensions; however, add the element of depth and photography becomes a little harder. The ease and the cost effectiveness of digital imaging means that once the initial equipment has been purchased, 3D photography becomes cheaper and more cost effective.

The simplest way into 3D photography is to buy a pair of digital cameras and join them together in the manner of the simple "disposable camera" stereo camera. Of course, with a flash digital camera, just using sticky tape is a bit too Heath Robinson; however, a more sturdy solution can be fashioned, using, for example, the tripod mounting screws shown earlier in the book in Fig. 3-28. Some simple construction should ensure that two cameras can be rigidly joined together for this type of work. However, this still poses some problems because both shutters must be fired simultaneously.

Method 2

For some years now, hobbyists have experimented with their own stereo digital cameras, taking existing camera bodies that are on the market and joining them together to form a stereo camera. There are a plethora of devices available for coupling shutter releases together, but up until now, no manufacturer has taken the initiative to release an off-the-shelf stereo camera. In an exciting development, Fujifilm has decided to release a stereo digital camera containing two lenses in a compact camera format (Fig. 15-10).

Figure 15-10   *Fujifilm Fine Pix Real 3D W1.*

The camera shoots native stereo pictures. What makes this package exciting is that Fuji have also produced a "digital photo frame" that comes with a lenticular screen in front of it. This digital photo frame enables the pictures taken using the stereo camera to be "replayed" in 3D that can be free-viewed.

Method 3

New on the market is the Minoru 3D Webcam, a device that allows you to see the world in 3D. Unfortunately, with a webcam, you are tethered to your PC. However, don't let this stop you. Any intrepid evil genius armed with a laptop should be able to overcome the constraints of the webcam's wire! The great thing about the Minoru webcam is to find a buddy who also has one. You can Skype and MSN in glorious three dimensions!

All of these different stereo imaging technologies produce a pair of images that will resemble Fig. 15-11, a view of the Empire State Building, taken from the "Top of the Rock." We've looked at the equipment we can use to view this

Figure 15-11   *An example of a stereo pair.*

stereo pair and looked at free-viewing methods, but what about if we want a 3D image that we can view without special equipment or spending prolonged periods trying to align and focus our eyes to the image?

Let's look at a couple of different projects that we can tackle once we have produced some stereo images to view them in different ways. As we've seen, holography is one technology; another is *lenticular imaging*.

You Will Need

- Computer
- Color printer
- Lenticular lens sheet (see text)

Tools

- Lentikit software (see below)

Have you ever seen a color print with vertical "lines" of plastic running through it that appears to be three dimensional?

Lenticular images are known as *autostereoscopic*, which means you can view them with the naked eye without any additional apparatus. But how does this work?

Lenticular prints consist of a specially printed image, where slices from a number of images are interleaved, laminated onto a sheet of plastic lenses that select one set of images for the eye to see depending on the viewing angle. This can be used to produce images with a 3D effect that do not require additional equipment to view.

Imagine a static object. If we look at it from a number of different viewpoints and take a picture,

we'll see a number of different images. Lenticular prints give us the ability to go beyond the "binocular" stereo view, taking pictures on either side of this, so that we can capture some of the horizontal parallax of the scene. This isn't quite as good as holography's ability to capture a scene from all angles at the same time, but it is an improvement on basic stereo imaging. Take a look at Fig. 15-12 to see what I mean.

Then we take these four images and slice them into stripes. Using a piece of software, we can then take these stripes and merge them into a single image, repeating stripes from the adjacent photographs in sequence. So with our four images, we would have a pattern of stripes on the final photograph that goes 1,2,3,4,1,2,3,4,1,2,3,4 . . . you get the picture! Or should I say, lenticular picture. Figure 15-13 should clarify.

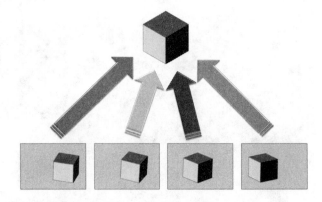

Figure 15-12 *A lenticular picture if comprised of two or more pictures taken from different horizontally transposed viewpoints.*

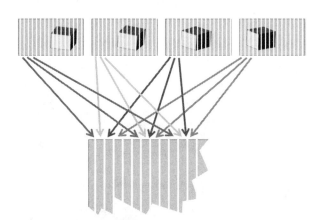

Figure 15-13 *Several pictures are chopped into lines, and then these lines are interleaved.*

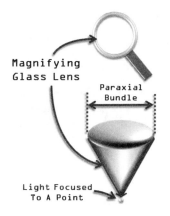

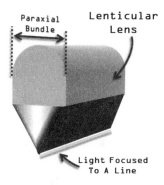

Figure 15-14 *Normal lenses focus to a point; lenticular lenses focus to a line.*

Now we need to understand how the lenticular lens works. Imagine a magnifying glass. A magnifying glass can be used to focus the sun, for example, into a point (and burn things!). However, the pattern that emerges from the other side of a magnifying glass is like a cone because the lens is circular. Lenticular lenses take the form of a stripe, whose cross section is kind of semicircular. The pattern that would emerge if we shone light through the "wrong" side of the lenticular sheet is a kind of triangular prism (i.e., it focuses the light to a line). Take a look at Fig. 15-14.

This means that if we have the lens focused on a line and look at the line from the other side of the lens, the narrow line will appear to fill the width of that lens. Now if we change our position horizontally, the line will move to an adjacent line. This is how the lenticular lens helps to "select" the right images for our left and right eyes.

As the viewer changes position, a different "stripe" is brought into focus, and because it is magnified it appears to fill the whole width of the lenticule. This is why we see a seamless image rather than stripes of a number of different images.

Now that we understand how the lens works, it's important to realize that all lenticular lenses are not the same. There are some crucial dimensions to the lens that affect how it distorts the image. We can see these in Fig. 15-15. For our purposes, we

should note that the most important one of these is the "period." When we buy a lenticular sheet, it will come with a measurement, commonly given in LPI, which is to say, lines or lenses per inch. The higher the LPI, the shorter the period (or finer the lens if you prefer). When we come to making our

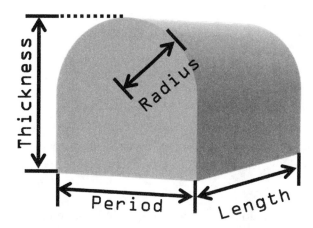

Figure 15-15 *Some key measurements of lenticular lenses.*

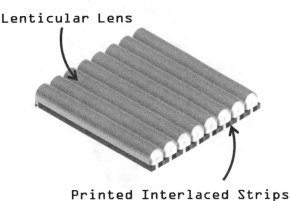

Lenticular Lens

Printed Interlaced Strips

Figure 15-16 *The construction of a lenticular lens.*

Figure 15-17 *The left image: "empire state left.jpg."*

own lenticular images, we need to ensure that the settings in the software correspond to the type of lens we have bought!

Now all that remains for us to do is align the image that has been printed behind a sheet of lenticular lenses, the construction of which is shown in Fig. 15-16. When the alignment is correct, the 3D effect should emerge! Let's try this for real!

How to Make a Lenticular Image

The open source community; or more specifically Andrew Rowbottom, has developed a fantastic application, Lentikit, which can be used to produce a lenticular image. Although the software is currently in alpha testing, it is quite usable.

Lentikit can be downloaded from Sourceforge at http://lentikit.sourceforge.net/index.html.

We're going to be working with the stereo pair of the Empire State Building, which has been taken into a simple photo editing program and

cropped into two separate images: the left image, whose file name is "empire state left.jpg" and the right image, which is called "empire state right.jpg," rather unsurprisingly!

We're going to use these images for the next couple of exercises: Fig. 15-17, the left image, and Fig. 15-18, the right one. These images were scanned from a photo taken with a Loreo stereo camera. You can see to the right of the left image and to the left of the right image a little characteristic

Figure 15-18 *The right image: "empire state right.jpg."*

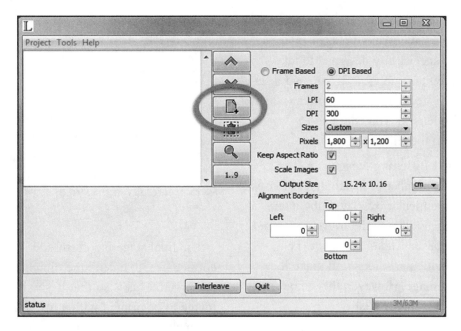

Figure 15-19 *The add files icon.*

shadowing from the way the camera is designed—with a barrier between the two halves of the frame. The two images are on the Evil Genius Web site, www.holographyprojectsfortheevilgenius.com, if you want to try out this project without needing to generate your own.

Install and open the Lentikit software. The first step is to add some images. These will be used to create the lenticular image we are going to generate. The icon you are looking for is shown in Fig. 15-19 and looks like a sheet of paper with a folded corner and a small + symbol. Click this to bring up a dialog box that allows you to select the files for your left and right images.

You will see that in the white box we have now added the two files to the program. This can be seen in Fig. 15-20, which also illustrates the next step, to select the "Frame Based" radio box in the

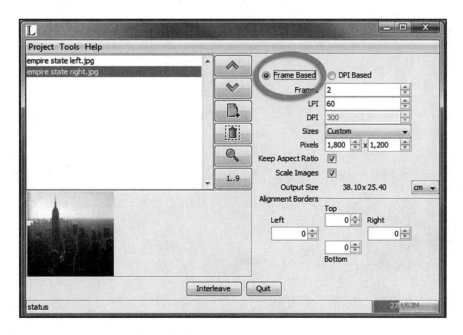

Figure 15-20 *Select the "Frame Based" radio box.*

software. We're going to be working with "two frames" because we have a stereo pair of images. Some cameras are specifically designed to take pictures from more than one view. If you had a set of processed images from a Nimslo camera (which has more than two lenses), you could add additional frames to the lenticular image.

Take note of the LPI setting, which is a measurement of the number of plastic linear lenses per inch that your lenticular overlay will contain. You will need to match this to the overlay material that you have purchased. Using a coarser (i.e., smaller number) lenticular screen will make it easier to align the image. However, this comes at the expense of poorer image quality.

Click the "interleave" button, highlighted in Fig. 15-21, and the two images are fused into a single image composed of multiple vertical slices of each image. Viewing this on its own, it appears a little disjointed; however, laminate it together with some lenticular film, and the lenticular lenses will direct each eye to the appropriate image and make it "jump" out from the page.

Hint

Figure 15-22 shows the final image with the interleaved vertical stripes. You can see how the two images have been composited into a single image if you get a magnifying glass or loupe and look carefully at the stripes.

The final stage is to print the image at the correct size, to ensure the dimensions are preserved. This is so that if you have generated an image to be printed with a resolution of 300 dots per inch, to be affixed to a lenticular sheet with a lens pitch of 60 lines per inch, the lens lines and image lines will align.

If your sheet of lenticular material does not come coated with its own adhesive, you should be able to offer it up to the printed image and see the lenses align. Try moving the sheet to the left and the right and seeing how the sheet lines up with the

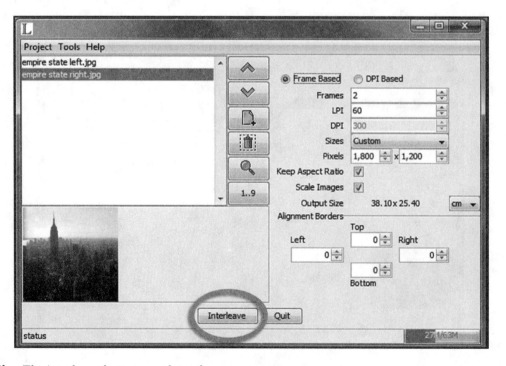

Figure 15-21   *The interleave button completes the process.*

Figure 15-22 *The final lenticular image ready for overlay with a lenticular sheet.*

image to best produce a good-quality 3D image. You can then use an adhesive that is optically clear and pressure sensitive to adhere your image to the lens sheet. There are some lens sheets available that are self-adhesive and simply require a backing

paper to be removed from the sticky surface before joining together.

Film-based lenticular imaging seems to have died in the past few years. There were a limited number of companies that would offer to process lenticular film, and it appears this has dwindled to none, with only specialist imaging companies able to produce lenticular images as a one-off service. However, with the stunning news of Fujifilm's 3D digital camera, and the growing interest in 3D imaging, I would be really unsurprised if in the next decade, a manufacturer doesn't emerge onto the market with a digital "lenticular camera" with multiple lenses. At first this might be used with an LCD screen with a lenticular overlay. But with cheap home printing technology, I think the dream is in sight of automated home lenticular printing.

Here are some sources of lenticular lenses:

http://www.3dphotopro.com/lenses.html

http://www.dplenticular.com/en/web/small_quantities/id_163

http://www.depthography.com/screen.html

Project 40: Make an Anaglyph Image

You Will Need

- Anaglyph maker software
- A stereo pair image
- Anaglyph glasses

This project uses a piece of freeware software that was developed by Takashi Sekitani from http://www.stereoeye.jp/software/index_e.html.

Another type of 3D image that you may have come across is the anaglyph. This is the type of image that at first glance looks like a cheap comic book that has

been badly printed with its colors out of register. Certain parts of the image have "red" fringes around them; other parts of the image have blue or green fringes around them. Upon closer inspection, we can see that it isn't a fault of the printer at all, but the different colors represent two different images. By using a pair of glasses with colored lenses, we can separate these images into "left" and "right" eye images. If you have a 3D anaglyph image on hand, then use this, or else browse onto www.holographyprojectsfortheevilgenius.com to view a color version of Fig. 15-26 later in this chapter.

You're going to need a pair of "anaglyph glasses" to view your anaglyphs, but the good

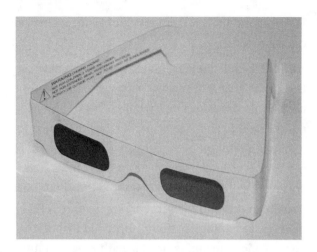

Figure 15-23 *Anaglyph glasses.*

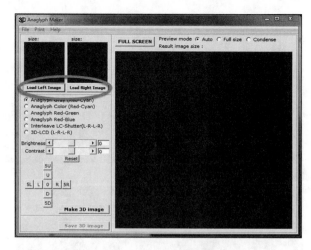

Figure 15-24 *Screenshot of Anaglyph Maker software with load image buttons highlighted.*

news is, that apart from that, you pretty much have everything you need with a standard desktop PC setup. Anaglyph glasses, as shown in Fig. 15-23, are available in a range of different colors; you will have a pair of lenses usually in red and another color of a bluish-green hue. The software will generate images for red-green, red-cyan, and red-blue glasses.

Furthermore, if you are using red-cyan glasses, as we will in this project, you have the option to print your image as either a "gray" image, where the two images are first converted into gray scales, and then toned in red and cyan for each respective eye, or you can produce a color image. Because you will be viewing the image through colored lenses, there will be an obvious loss of color information over, say, viewing a color photograph. However, in this mode, the program attempts to tone the image in such a way that color information is retained for portions of the image where the overlapping image is the same, using only red and cyan for the parts of the image that are different.

Anaglyph images use color to separate the left and the right eye images, so with a color printer and desktop PC and one of the camera solutions listed in the early part of this chapter, you have everything you need to start producing anaglyph images.

Fire up your PC, and download and install the anaglyph maker software. The interface is very easy to use.

There are two separate buttons for "Load Left Image" and "Load Right Image"; click each of these and select your two files in turn. We've used the files with the pictures shown in Figs. 15-17 and 15-18. You can see the screenshot of the software in Fig. 15-24. There are then a number of different radio buttons that you can select depending on the type of 3D image you want to generate. We're going for a red/cyan-gray image.

There are adjustments that you can make to the brightness and contrast of the image. Below these adjustments you will see a directional arrow that can be used to aid with the alignment of two images. Click the "Make 3D image" button highlighted in Fig. 15-25.

The software will generate an anaglyph for you. We can see in Fig. 15-26 how the software shows you a preview of the final image.

I've overlaid the text because the color information is lost in this black-and-white book. When you save the file, you will get something that looks like Fig. 15-27, only in color!

Although we can't see the color information, and if you tried to view this image using anaglyph glasses, you would see no depth to the image. But

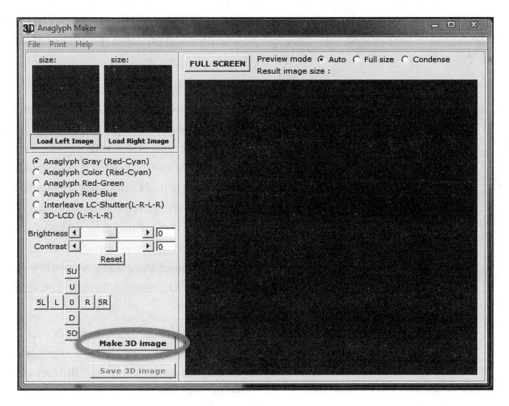

Figure 15-25   *The "Make 3D Image" button.*

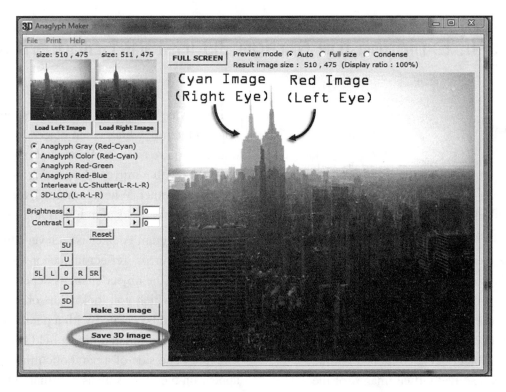

Figure 15-26   *The image being previewed in the software.*

Figure 15-27 *A black-and-white print of a color anaglyph (check the Web site for a color version).*

we can see how the technique differs from the lenticular image. This time, the image isn't divided into vertical lines, but the two images are "ghosted" together, one image blended into another. The

effect will become clearer if you log on to www.holographyprojectsfortheevilgenius.com and view the color version with anaglyph glasses!

We've only covered a smattering of 3D imaging technologies, but you should now have a sound understanding of the ones you are most likely to come into contact with. The great thing about technology is that it develops so fast. As we have seen, 3D imaging has been around for ages, but it's never really taken off. With the advent of digital imaging and the ability to transfer imaging information at the speed of light, without the need for a bulky physical product, 3D imaging has the capability to come into its own.

The last few projects have focused on producing flat images that can be printed out and handled much like a flat hologram. But before I go, I'll leave you with one more nice little gizmo, which although mistakenly billed as an "instant hologram maker" (which it isn't!) is sufficiently interesting to warrant a mention.

Project 41: Exploring the Mirage 3D Instant Hologram Maker

You Will Need

- Mirage Instant Hologram Maker
- Small object to place inside hologram maker

Let's state clearly from the beginning that although this is called a hologram maker, the image that this device produces definitely isn't a hologram. It doesn't record the image in a form where it can be replayed. This device is wholly reliant on the object being contained within the device to produce the "virtual image" of the object.

The device consists of two mirrors, one with a hole on the top, which when placed face each other. If you're a fan of Asian cooking, imagine two woks with a mirror-like coating on their inside facing each other. When combined, they make something that looks akin to a flying saucer.

The "hologram" maker is a clever illusion, where two mirrors reflect in such a way that the reflection of the object changes depending on the angle from which you look at the object. A simple illusion could be created with a parabolic mirror, whereby an illuminated object held behind a screen on one side of a parabolic mirror would appear on the other side as a virtual 3D "real image."

We must be a little bit careful about terminology. Correctly, "virtual image" means an image where light rays appear to come from a reflected image but do not. It is virtual images that we see in our holograms, where the actual object has been removed, but the hologram still manages to recreate a 3D wavefront. By contrast, a real image is one where the reflected light rays actually come from the object itself, and this is the case with the Mirage 3D Instant Hologram Maker.

Look at Fig. 15-28. The image appears to be of a strawberry sitting on top of a mirror, however, that strawberry's position in 3D space appears to be

Figure 15-29 *The path of the light rays inside the Mirage Instant 3D Hologram Maker.*

Figure 15-28 *The Mirage 3D Instant Hologram Maker in action.*

above the "flying saucer" or "upturned wok" depending on your preferred imagery. In fact, the strawberry is sitting inside of the "flying saucer." You are looking at the bottom parabolic mirror, which in turn is facing the top parabolic mirror, which in turn is facing the object. Figure 15-29 should go some way toward illustrating the paths of the light rays and helping you to understand how this illusion works.

This is very clever. It certainly causes you to pause for a moment and consider the dynamics of how it works. However, in reality, we know that however finely manufactured, this device is just a pair of mirrors cleverly shaped to make an object seem in a different place from where it is, a cheap conjuring trick really compared to the true magic of holography, a miracle that allows the light fronts of objects to be recreated anywhere *even when they aren't there!* Despite this, I still think it is a nice novelty and worth experimenting with!

Chapter 16

What Next for Holography?

As we have seen in this book so far, there are already a whole stack of applications for holography in the present day. But with the technology rapidly advancing, let's look ahead to the applications of holography in this century.

Data Storage

We are already familiar with the use of data storage based optical media in the CDs, DVDs, and blue-ray discs on a regular basis to watch films, browse multimedia content, and play computer games.

However, as we have seen, holographic technology allows us to store an immense amount of information on a two-dimensional plane. Figure 16-1 shows how Holographic Versatile Discs (HVD) could read and write information using a novel arrangement of lasers to both read and write data and address that data. Could holography allow us to create discs the size of present-day DVDs that allow us to store hundreds

HVD 1–10TB

Blu Ray 25–100GB

DVD 4.7–17.08GB

CD Up to 700MB

Figure 16-2   *How the capacity of HVD's compare to other optical disc technologies.*

of gigabytes, even a terabyte, of information on a single disc? Figure 16-2 shows how HVD's have the potential to store much more information than their other optical disc counterparts; in fact HVD's have the potential to store up to a massive 6TB of information! We've seen with holograms that every part of the hologram contains data about the whole, and if we break a hologram, we can still see the whole image, albeit from a limited perspective. Could this integral feature of holographic technology provide redundancy and data security? If your disc gets scratched, it continues to work, just that the reader has to "look" at the data from a different angle and compute what is behind the scratch.

Imagine a disc with 4000 h of video viewing time. The HVD, holographic versatile disc, is already under development. Potentially, when the technology is fully realized, a disc the size of present CDs or DVDs could hold up to 3.9 terabytes of information. What's more, an HVD

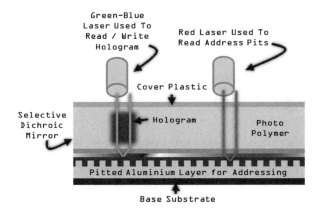

Figure 16-1   *The technology used to read Holographic Versatile Discs.*

system would be able to transfer data incredibly quickly. The sequential "bit by bit" reading of DVD and CD technology is dispensed with, and data is read holographically—"massively" in parallel! Data transfer speeds of up to 1 gigabyte per second are thought possible. Whereas each time your DVD writers laser pulses on and off, a single bit is written to the disc. With HVDs, 60,000 bits of data would be written with that same pulse.

Dynamic Holography

Although the technology has been with us for a number of years to be able to produce moving pictures in three dimensions, the idea has never really taken off. A simple stereo film requires twice the amount of information as a conventional moving picture but produces a three-dimensional image from a fixed viewpoint. Although 3D movies have enjoyed limited success at cinemas, there has never been a great appetite for 3D in the home. Why is this? Well, the need for glasses, as is the case with many stereo television systems, is a bit of an encumbrance, one that a holographic technology could potentially solve. Some manufacturers have suggested the idea of using lenticular screens with several channels of video providing some scope for "free-viewing" the picture without the need for glasses. Screens of this type will probably be on the market in the next few years.

However; a true holographic screen that could provide not only three dimensionality but also multiple viewpoints would have the potential to find a home in many applications—not just for entertainment, but imagine the applications in other fields of technology: Expert surgeons from the other side of the planet able to offer assistance and aid to doctors performing surgery while being able to see the totality of the situation they are facing from all angles.

Undoubtedly holographic optical elements are going to find more and more applications. At the moment, they are used in applications as diverse as allowing multiple beams to scan the barcodes on your groceries, to providing "heads-up" displays for the pilots of fighter aircraft, and in that last application, you may see the same technology coming to a vehicle near you. With the abundance of in-car electronics, it's getting harder and harder to sort the important information out from the distractions. The cars of science fiction for many years have displayed data on the windshield of the vehicle, and some high-end vehicles already have this capability. But future vehicles might have advanced safety features, such as night vision cameras identifying hazards and projecting an identifying target on the windshield using a holographic optical element to inform the driver of danger ahead.

Security

At the moment; we see generic holograms used as security devices in credit cards and passports. However, as security specialists work to devise increasingly complex and cunning methods of preventing forgeries, the criminal underworld manages to catch up with the technology. What if tomorrow, though, rather than a generic hologram bearing the name of your bank or card provider, individual customized holograms adorned every piece of plastic in your wallet? Capturing a user's portrait in three dimensions would be a sure-fire way to reduce the possibility of fraud or identity theft in tomorrow's world.

Medicine

As medical imaging improves, doctors have access to an increasing array of information, which can help in the treatment of patients. However, in the future, rather than a handful of blurry black and white x-rays, might physicians not take out of a patient's medical notes some detailed crisp holograms generated from three-dimensional computed scans of the patient? Such rich information would doubtless aid complex diagnoses and surgery.

Novelty

At the moment; photo machines are commonly found in malls and stores across the globe. For a small fee, you can print and customize color photographs or stickers using a simple screen interface for a handful of coins. The technology is already within our grasp for commercial "holobooths" where rather than having to travel to a specialist's studio, you could sit in a booth at a mall of your choice, and within minutes a custom-printed photopolymer hologram of your face against whatever three-dimensional background you choose is produced!

Consumer Electronics

We've seen the boom in display devices in portable electronics. A decade ago, the thought of a phone with a color screen was still over the horizon, and the most sophisticated entertainment you could get on a mobile was a game of Snake. Today's electronic devices place entertainment in the palm of your hand, putting videos, games, and streaming videoconferencing at your fingertips. But what of tomorrow? We've already seen holographic elements used to produce a compact "virtual keyboard" whose keypad can be projected onto virtually any surface in a red laser glow. However, tomorrow we're likely to see pocket electronic

devices that through the miniaturization afforded by holographic optical elements are able to project a full-color picture, to allow your small device to produce a respectable-size screen output. Video projectors appear to shrink by the minute, and holographic optical elements will doubtless feature in the projectors of the future.

Furthermore, although videophones have been predicted by futurists for many years, it seems they have never really taken off. Might "holo phones" employing dynamic computer-generated holography unlock the door to more realistic virtual communication, reducing the need to travel across continents and oceans, by making the virtual face-to-face experience more realistic?

As technologists search for ever-faster processing power, might the future be optical rather than electronic, with holographic optical elements being used to help process data at ultrahigh speeds?

Documenting the Past . . . and the Future!

Holography has been used to make a record of artifacts and antiquities for future generations. Capturing something delicate or fragile in silver halide emulsion, the three-dimensional likeness of the object is captured forever.

Using the knowledge in this book, why not capture the things that are important to you today? After all, today's hobbyist holograms will be collectible three-dimensional glimpses of the past in the future!

The future of holography is in your hands.

Aberration: An optical distortion that causes an image to be unclear or unsharp.

Achromatic: Uncolored; a single color.

Amplitude: A word used when talking about waveforms, amplitude refers to the distance between the "peak" of a wave and a line running through the center of the waveform, which represents the mean of all values.

Anaglyph: A three-dimensional imaging technique (not holography!) that consist of two images, printed on top of each other; both in contrasting colors. These are viewed using "anaglyph glasses," which have matching lenses of contrasting colors. The lenses ensure each eye receives the correct image.

Angle of incidence: When talking about light falling on a flat surface, the angle of incidence refers to the angle between a line that points straight out from the surface of the plane, perpendicular to it, and the line formed by the "incident beam" or incoming light.

Animated hologram: A composite of several different images that have been separately recorded so that as the eye's viewpoint changes, the illusion of motion is created.

Beam ratio: The ratio between the relative intensities of the object beam and reference beam.

Beamsplitter: A beamsplitter takes a single laser light source and splits it into two distinct beams. This is really useful when we want to create a "reference beam" and an "object beam."

Coherent light: Light that is of the "same frequency" and also "in phase." Laser light is "coherent," whereas many light sources; for example light from an incandescent bulb, is not.

Computer-generated hologram: A hologram created using a three-dimensional model inside a computer and sophisticated software as opposed to being created using optical methods.

2D/3D hologram: A pseudo three-dimensional hologram made from many individual two-dimensional frames layered one on top of another.

Diffraction: As light passes through small openings, it diffracts or "bends."

Dennis Gabor: The Hungarian physicist who discovered "holography," he is considered the "father of holography: and received the Nobel Prize in 1971 for his work.

Developer: Chemical processing stage. When the film or plate is immersed in developer, the silver halide crystals that have been exposed to light are converted by the reducing action of the developer into silver, which appears black. This action turns the image from one that is weak and sensitive to light to one that is dense and no longer sensitive to light.

Embossed holograms: A hologram such as the one you have created using the processes in this book is turned into a metal replica that is a "master" stamp. This master stamp is then stamped very quickly onto a cheap substrate, such as metallized plastic Mylar, producing very cheap holograms.

Emulsion: The light-sensitive coating on which we produce holograms. It consists of light-sensitive silver halide crystals suspended in a binder that is coated onto the substrate whether it is film or plates.

Film: Usually refers to a flexible substrate that is coated with a light-sensitive emulsion. Film is not rigid and needs to be supported strongly throughout the exposure. If the film flexes, defects will result in the hologram. The alternative to film is "plates," which is photographic emulsion coated onto a rigid substrate.

Flip-flop holograms: As the viewer changes position, a flip-flop hologram alternates between two different images.

Full color: Simple holograms represent three-dimensional images monochromatically; however, it is also possible to have full-color holograms that can be produced using sophisticated optical or computer-generated methods.

Hologram: A three-dimensional image that is recorded on a two-dimensional plane. Holograms can be recorded optically on photographic emulsion or digitally using computer-generated imagery.

Holographic optical elements: A holographic reproduction of an optical element. Complex optics can be made cheaply into a holographic equivalent that will perform a similar job.

Holography: The art and science of recording three-dimensional images onto a two-dimensional surface is known as holography. It comes from the Greek roots *holos,* meaning "whole," and *gram,* meaning "message."

HOEs: See holographic optical elements.

Interference pattern: A record of the interference of the waves of two different beams of light: the "object" beam and the "reference" beam.

Inverse square law: A concept in physics, which states that from a point light source (where the light spreads out), illumination falling on a flat surface will be inversely proportional to the square of the distance from the point light source.

Lenticular imagery: A lenticular image consists of usually four or more pictures taken from slightly different viewpoints that have been chopped into pieces, interleaved, then displayed behind a lens which diverts the eye to the appropriate image depending on its angle of view.

Lenticular screen: A piece of plastic that is imprinted with cylindrical lenses, giving it a "lined" appearance. Lenticular lenses can be laminated to a specially printed piece of paper with a number of interlaced images printed at the same pitch as the lenses to create a three-dimensional image that does not require any apparatus to view. Images are in full color, and although sometimes mistaken for holograms due to the three-dimensional effect, they use completely different technology and cannot really be considered holograms.

Master hologram: When producing holograms commercially, a "master" hologram is created from which cheaper copies can be reproduced.

Multiplex holograms: Holograms composed of a number of still two-dimensional frames, which are integrated to form a single three-dimensional image that can move if required.

Nanometer: A millionth of a millimeter, and a measurement used when discussing the wavelengths of light. You will hear lasers described as producing light that is # # # nm or # # # nanometers in wavelength. This is to say that the distance from the peak of one wave to the next is # # # millionth of a millimeter!

Object beam: The beam of laser light that first hits the subject that is being recorded, before being reflected off this onto the holographic emulsion.

Opacity: The ability of a material to block the passage of light.

Optics: The science of understanding the properties and behavior of light and its interaction with different optical elements.

Overdevelopment: When a film or plate is overdeveloped, the optimum amount of development has been exceeded. This can result from exceeding the time recommended by the manufacturer when immersing the plate or film in chemicals, or it can arise from the chemicals used for development at a higher temperature than recommended. It can also be caused by mixing the chemical developer to the wrong strength. Overdeveloped images exhibit higher than usual density and contrast.

Overexposure: When a film or plate is overexposed, it has received excessive light as a result of the intensity of light, exposure time, or an incorrect combination of the two.

Photopolymer: A material that presents an alternative to traditional photographic emulsions. Rather than chemically reacting silver salts to produce areas of "clear" and "black," a photopolymer is a soft plastic that "cures" when exposed to light and is processed.

Photopolymer hologram: A hologram that is produced on a special recording material known as a "photopolymer." These holograms tend to have a green/amber tint to them and are very bright compared with many holograms.

Plates: A rigid substrate, usually glass, that is coated with a light-sensitive emulsion. Plates, although more expensive than film, is easier to work with for beginners.

Principle axis: When describing lenses, the principle axis can be thought of as an imaginary line that passes through the center of curvature of all lens elements.

Processing: A piece of holographic film is immersed in a sequence of different chemicals under controlled conditions. This turns the "latent" image formed during exposure into a permanent image that is durable and no longer light sensitive.

Rainbow holograms: A specific kind of hologram; as the viewer moves their head up and down, the images change through a variety of rainbow colors.

Reciprocity law: A law that can be used to understand the nature of a holographic exposure. Exposure = Time × Intensity; exposure is the sum amount of light that hits the emulsion given as an intensity of light over a period of time.

Reference beam: A beam that does not reflect off any object but hits the photosensitive emulsion directly. It is the interaction of the object beam with the reference beam that forms the pattern of light and dark recorded on the emulsion.

Reflected light: Light that does not pass through a subject but bounces off it (e.g., a mirror).

Reflection hologram: Hologram that requires the viewer to see light reflected from the surface of the hologram. The light source therefore must be on the same side as the viewer.

Refraction: As light passes through materials that are transparent but of different densities, it will change direction. You can see this effect if you put a pencil in a glass of water.

Safelight: A light that is selected to be of a color and intensity that it will not affect photographic materials that have been sensitized to respond to light from a narrow band of wavelengths.

Shutter: A device that is used to prevent light passing. Shutters are used to ensure that exposures are timed so that the correct amount of light reaches the photosensitive material.

Silver halides: Chemical crystals used in photographic and holographic light-sensitive emulsions. They react to light by changing from their clear state to a black state when exposed to light and, correctly processed, form a negative image.

Squeegee: A tool that comprises a pair of "tongs" with rubber blades or rollers on each tong. Squeezing the tong and wiping over a holographic plate or film will remove water from the surface of the hologram once it has been thoroughly washed.

Stereo camera: A camera that is designed to take two images of an object from the perspective of two points that are the same distance apart as the human eyes. These images can then be viewed to produce a three-dimensional image from a single point of view. However, unlike a hologram, the scene will not change when the viewer's position changes.

Stereoscopy: The art and science of creating the illusion of three dimensions using a pair of two-dimensional images. Stereoscopy is *not* the same as holography.

Test strip: A method of estimating the correct exposure time for a photosensitive plate by uncovering the plate in several stages over a timed period. This approach can be a little tricky with holography because any vibration of the film can ruin the resulting image.

Transmission hologram: A reflection hologram is one that requires the viewer to see light that passes through the hologram. The light source therefore must be on the opposite side to the viewer.

Transmitted light: Light that passes through a film, plate, or object that is transparent or translucent. The optical density of the material that the light is passing through and the intensity of the light will determine the amount of light that is "lost" as it passes through.

Underdevelopment: When an image is not developed to the correct degree as a result of shortening its development time, reducing the temperature of the chemicals used to process it, or using a working solution of the wrong concentration, it results in an image with poor contrast and less density than a correctly developed image.

Underexposure: A condition that arises when insufficient light reaches the film.

Volume hologram: A hologram where the angle between the object beam and the reference beam is obtuse.

Washing: The last stage of the development process, which helps to remove any residues of chemicals that are formed during the processing cycle and stick to the film or plate.

Water bath: To keep photo-processing chemicals at the correct temperature, it is sometimes necessary to immerse them in a "water bath," which helps to keep them at the correct temperature required for development. By adding cool or warm water, the temperature can be adjusted to the optimum for the process.

Wavelength: The distance from the peak of one wave to the peak of the next. The wavelength of light corresponds directly to the color that we perceive it to be.

Wet processing: A system where chemicals in fluid form, mixed with water to the required concentration, are used to turn the latent image on the film into a permanent image that can be viewed safely in ordinary light.

White light: Unlike laser light (which is "coherent" and contains a single wavelength of light), white light contains multiple wavelengths of light and thus is "incoherent."

Working solution: A chemical that has been mixed to the correct concentration for use in processing holograms.

Suppliers' Index

North America

Integraf LLC

218 Main Street #674

Kirkland, WA 98033 USA

Tel. (425) 821-0772

Tel. (425) 821-0773

E-mail: info@integraf.com

http://www.holokits.com/

Integraf sells a range of plates, film, and holographic chemicals and produces a fantastic range of kits to suit a range of wallets, all supplied with very clear instructions.

Industrial Fiber Optics, Inc.

1725 West 1st Street

Tempe, AZ 85281-7622 USA

Tel. (480) 804-1227

Tel. (480) 804-1229

E-mail: INFO@i-fiberoptics.com

Industrial Fiber Optics sells a range of laser education kits. I particularly recommend the "Sandbox Holography Kit" as being really well put together and easy to use. For a sensible price, it takes the hassle out of fabricating your own components.

Laser Reflections

P.O. Box 655

Seabeck, WA 98380 USA

Tel. (360) 830-5936

E-mail: info@laserreflections.com

http://www.laserreflections.com/

Suppliers of holographic film and plates

Sterling Resale Optics

E-mail: Sterling532@aol.com

Suppliers of a wide variety of optics.

Laser Resale Inc.

54 Balcom Road

Sudbury, MA 01776 USA

Tel. (978) 443-8484

Fax: (978) 443-7620

E-mail: laseresale@aol.com

http://www.laserresale.com/

Supplier of a wide range of lasers.

Thorlabs - Newton, New Jersey

435 Route 206 North

Newton, NJ 07860 USA

Tel. (973) 579-7227

Fax: (973) 300-3600

E-mail: sales@thorlabs.com

http://www.thorlabs.com/

Suppliers of a wide range of professional optical equipment, optical benches, and optic components.

Forth Dimension Holographics Inc.

PO Box 259

Nashville, Indiana 47448 USA

Tel. (812) 340-9161

E-mail: info@forthdimension.net

Suppliers of Slavich holographic emulsions.

Photographers' Formulary, Inc.

P.O. Box 950, 7079 Hwy 83 N

Condon, MT 59826-0950 USA

Tel. (800) 922-5255 (toll free)

Tel. (406) 754-2891 (toll call)

Fax: (406) 754-2896

E-mail: formulary@blackfoot.net

http://www.photoformulary.com/

Purveyors of unusual photographic chemicals and
formulations.

VinTeq, Ltd.

8537 Purnell Ridge Road

Wake Forest, NC 27587-7557 USA

Tel. (888)556-9924 (toll free)

Tel. (919) 556-9924 (toll call)

Fax: (919) 556-9992

E-mail: vinson@vinteq.com

Europe

Geola Digital Uab

41 Naugarduko

PO Box 343

LTU-03227 Vilnius, Lithuania

Tel. +370 5 2132 737

Fax: +370 5 2132 838

E-mail: info@geola.com

http://www.geola.com/

Lasers and holographic emulsions.

Thorlabs GmbH

Hans-Boeckler-Str. 6

85221 Dachau/Munich

Germany

Tel. +49 (0) 8131-59-56-0

Fax: +49 (0) 8131-59- 56-99

E-mail: europe@thorlabs.com

Suppliers of a wide range of professional optical
equipment, optical benches, and optic
components.

Thorlabs, Ltd.—Cambridgeshire, United Kingdom

1 Saint Thomas Place, Ely

Cambridgeshire CB7 4EX

UK

Tel. +44-1353-654440

E-mail: sales.uk@thorlabs.com

Suppliers of a wide range of professional optical
equipment, optical benches, and optic
components.

Colour Holographic

Unit L Trinity Buoy Wharf

64 Orchard Place

London E14 0JW

Tel. +44 (0) 207 494 8250

E-mail: jonathanlewis@colourholographic.com

http://www.colourholographic.com/

Professional holography and color holographic
plates.

Index

Note: Page numbers referencing figures are followed by an "*f*".

D

E

F